Energy

Editor: Danielle Lobban

Volume 413

First published by Independence Educational Publishers

The Studio, High Green

Great Shelford

Cambridge CB22 5EG

England

© Independence 2022

Copyright

This book is sold subject to the condition that it shall not, by way of trade or otherwise, be lent, resold, hired out or otherwise circulated in any form of binding or cover other than that in which it is published without the publisher's prior consent.

Photocopy licence

The material in this book is protected by copyright. However, the purchaser is free to make multiple copies of particular articles for instructional purposes for immediate use within the purchasing institution. Making copies of the entire book is not permitted.

ISBN-13: 978 1 86168 872 9

Printed in Great Britain

Zenith Print Group

Contents

Chapter 1: About Energy
A complete guide to renewable energy vs fossil fuels	1
Top 5 fastest-growing renewable energy sources around the world	5
Hornsea 2, the world's largest windfarm, enters full operation	6
How do UK energy prices compare with the rest of Europe?	7
How much of the UK's energy is renewable?	8
Why the UK's unfair energy market is unlikely to spearhead a green transition	10
Rising sales of electric vehicles show they are here to stay	13
Electric cars are greener than petrol cars – but they're far from perfect	15
Are smart solar flowers worth the recent hype?	16
High fossil fuel prices are good for the planet – here's how to keep them high while avoiding riots or hurting the poor	18
A town in Devon now gets its gas supply from animal poo	19

Chapter 2: Energy Crisis
Why is there an energy crisis?	20
British Gas owner's profits increase five-fold while energy bills soar	21
How energy retail suppliers are supporting customers through the crisis	22
In an energy crisis, every watt counts. So yes, turning off your dishwasher can make a difference	24
Martin Lewis 'ringing the alarm bell' on energy crisis as he receives CBE	26
Facing up to the global energy crisis	27
Breaking the gas habit: the energy crisis spotlights the need for green alternatives	28
Building nuclear power stations in Scotland will help solve energy crisis, says Sir Keir Starmer	29

Chapter 3: Future of Energy
What the future of renewable energy looks like	30
World's first sand-powered battery is ready to heat homes	32
Shell to build Europe's biggest renewable hydrogen plant	33
The English city using community energy to drive positive change	34
Renewable energy –powering a safer future	36
Powered by poo? The weird and wonderful alternative energy sources that could soon be powering you home	37

Key Facts	40
Glossary	41
Activities	42
Index	43
Acknowledgements	44

Introduction

Energy is Volume 413 in the **issues** series. The aim of the series is to offer current, diverse information about important issues in our world, from a UK perspective.

ABOUT ENERGY

We are in the midst of an unprecedented global energy crisis caused by soaring fuel prices. The need to harness affordable, reliable and sustainable power sources has never been more urgent. This book explores the different types of renewable energy currently in use, the efforts to cope with the energy crisis and the ongoing development of alternative clean, green energy sources to tackle climate change.

OUR SOURCES

Titles in the **issues** series are designed to function as educational resource books, providing a balanced overview of a specific subject.

The information in our books is comprised of facts, articles and opinions from many different sources, including:

- Newspaper reports and opinion pieces
- Website factsheets
- Magazine and journal articles
- Statistics and surveys
- Government reports
- Literature from special interest groups.

A NOTE ON CRITICAL EVALUATION

Because the information reprinted here is from a number of different sources, readers should bear in mind the origin of the text and whether the source is likely to have a particular bias when presenting information (or when conducting their research). It is hoped that, as you read about the many aspects of the issues explored in this book, you will critically evaluate the information presented.

It is important that you decide whether you are being presented with facts or opinions. Does the writer give a biased or unbiased report? If an opinion is being expressed, do you agree with the writer? Is there potential bias to the 'facts' or statistics behind an article?

ASSIGNMENTS

In the back of this book, you will find a selection of assignments designed to help you engage with the articles you have been reading and to explore your own opinions. Some tasks will take longer than others and there is a mixture of design, writing and research-based activities that you can complete alone or in a group.

FURTHER RESEARCH

At the end of each article we have listed its source and a website that you can visit if you would like to conduct your own research. Please remember to critically evaluate any sources that you consult and consider whether the information you are viewing is accurate and unbiased.

Useful Websites

www.earth.org

www.energy-uk.org.uk

www.friendsoftheearth.uk

www.greenpeace.org.uk

www.hornseaprojects.co.uk

www.independent.co.uk

www.inews.co.uk

www.kcl.ac.uk

wwww.metro.co.uk

www.money.co.uk

www.nationalgrid.com

www.nesta.org.uk

www.news-decoder.com

www.positive.news

www.telegraph.co.uk

www.theconversation.com

www.theecoexperts.co.uk

www.thisismoney.co.uk

www.un.org

Chapter 1: About Energy

A complete guide to renewable energy vs fossil fuels

By Josh Jackman

Goodbye, oil. Farewell, natural gas. So long and don't bother to write, coal. The world is increasingly patting fossil fuels on the back, then sending them off to the great sooty place in the sky.

As humans face up to the catastrophic consequences of climate change, people have finally woken up to the desperate need to banish these dangerous energy sources in favour of renewable energy.

So far, 185 countries have signed the 2016 Paris Agreement, and the results have been clear. Governments all over the world have taken more and more actions to cut their fossil fuel usage to meet (or even exceed) the treaty's requirements. One consequence is that one third of global power capacity is now based on renewable energy, according to a report in April 2019 from the International Renewable Energy Agency – and that's only the start.

France, Britain, and Germany (among others) have committed to getting rid of all their coal-fired plants by 2021, 2025, and 2038 respectively. An article in The Financial Times in 2019 questioned the value of coal power by asking: 'Why invest in capacity that has a 20-year lifespan at most?' It's a good question.

The amount of renewable energy produced in the European Union increased by two thirds between 2007 and 2017, while consumption more than doubled between 2004 and 2017. Clearly, renewable energy is on the rise.

The list goes on, with Costa Rica's president stating his country would stop using fossil fuels by 2050, Norway instituting a ban on the sale of petrol and diesel-powered cars by 2025, and the Netherlands doing the same by 2030. Furthermore, since the start of 2017, Dutch national rail network trains have been completely powered by renewable energy.

Over in the US, Nevada, California, Hawaii, New Mexico, and Puerto Rico have all committed to using only renewable energy by 2050, along with more than 100 cities and towns across the country. This switch makes economic sense, too, as 74% of American coal production now costs more to produce than solar and wind energy, according to a 2019 study by renewables analysis firm Energy Innovation. By 2025, this figure is estimated to rise to 86%.

Coal in particular is dropping off the face of the energy map, generating just 5% of the UK's power in 2018. The British government reported at the end of the year that the output from bioenergy, waste, wind, solar, and hydro was around 13 times higher than coal. For more than two weeks in summer 2019, Britain didn't use any coal to generate electricity – the longest such period since 1882.

Fossil fuels are disappearing, while renewable sources rise to prominence. But let's slow down and answer some commonly asked questions. For instance, what exactly is renewable energy? What's the importance of renewable energy? And how do you define non-renewable energy sources? Read on to discover the answers to all your questions.

Did you know?

According to NASA, humans have increased atmospheric CO_2 concentration by more than a third since the Industrial Revolution began.

What is renewable energy?

Renewable energy is any energy which comes from a source that regenerates on a human timescale (a matter of minutes or months, rather than millions of years) and produces minimal emissions that contribute to climate change. Natural phenomena like the sun, wind, water, and natural heat stored in the Earth (known as geothermal energy) are prime examples of renewable energy.

Simply put, renewable energy is the future.

Renewable energy sources

Wind power: Air flow moves wind turbines, creating mechanical power from kinetic energy, which a generator then converts into electricity.

Solar power: Energy from the sun's light is converted into electricity.

Hydro power: Energy derived from the fast flow of water, generally through spinning a turbine, which a generator can then convert into electricity.

Tidal power: Using the movements of the tides to generate electricity.

Geothermal energy: This is thermal energy derived from deep in the Earth, coming from radioactive decay and the continual heat loss which began when the planet was formed.

Ambient heat captured by heat pumps: Pumps take heat from the air or ground and use it to heat a building.

Biofuels: Fuel made with crops or biological waste.

Waste: This involves turning non-biological waste into energy, usually through burning it.

What are fossil fuels?

Fossil fuels are non-renewable energy sources, namely coal, oil, and natural gas. They were created tens or hundreds of millions of years ago, when plants and animals were buried and compressed into kerogen. This mineral, which contains 10,000 times more organic matter than any living being, is the basis for all fossil fuels.

Over the course of millions of years, geothermal heat turns kerogen into one of coal, crude oil, or natural gas.

Fossil fuels still make up around 80% of the world's energy consumption, but are considered non-renewable since it takes such a long time to produce them. As well as their effectively finite nature, they're also being phased out because of their negative effects on the Earth's climate. According to NASA, humans have increased atmospheric CO_2 concentration by more than a third since the Industrial Revolution began.

Non-renewable energy sources

Coal: A carbon-rich substance which comes from dead plants decaying, turning into peat, and being converted through millions of years of heat and pressure into coal. It is extracted from inside the Earth and burned for energy. In 2011, coal plants produced 14.4 billion tons of CO_2.

Crude oil: This is created when lots of tiny dead animals are buried and exposed to intense heat and pressure over millions of years. Like coal, it is then burned for energy. Crude oil can be turned into refined petroleum, kerosene, orimulsion, and propane.

Natural gas: This is made from layers of decaying animals and plants being subjected to intense heat and pressure for millions of years. It is then burned for energy.

What is the importance of renewable energy?

Renewable energy sources will help to slow down and prevent the worst consequences of climate change. In effect, they can help save the world from the destructive practices humans have undertaken since the Industrial Revolution began more than two centuries ago.

NASA has reported that the intensity, frequency, and duration of North Atlantic hurricanes 'have all increased since the early 1980s', and that 'storm intensity and rainfall rates are projected to increase as the climate continues to warm.'

The organisation has also predicted that 'in the next several decades, storm surges and high tides could combine with sea level rise and land subsidence to further increase flooding in many regions.'

The World Wide Fund for Nature has stated that climate change is 'likely to be the greatest cause of species extinctions this century'. This has been highlighted by a United Nations body's report which claims a 1.5°C average rise may put 20-30% of species at risk of extinction.

'If the planet warms by more than 2°C, most ecosystems will struggle.'

Even more worryingly, a 2017 University of Washington study based on the data in this UN report found there was a 90% chance this would happen – so it's past time that we embrace renewable energy.

Non-renewable energy sources are also finite, meaning that at some point in the next century or so, they will run out – leaving us with a huge power deficit. If we're going to carry on powering everything from your smartphone and the lights in your home to planes, trains and automobiles, we need renewable energy sources.

Soon, alternative energy won't just be the better option; it'll be the only option.

Pros and cons of renewable energy

Pros:
- Will help to save human and animal lives
- Will stop deadly pollution
- Stable energy prices
- A practically endless supply
- Job creation
- Lower running costs
- Can be produced on a smaller scale
- If a solar panel or wind turbine malfunctions, it doesn't lead to mass deaths or illnesses

Cons:
- Some types are currently more expensive
- Inconvenient to overhaul the energy sector
- Can require larger areas of land
- Some rely heavily on a climate that's changing
- Staff would need to be retrained or replaced

Information updated in July 2019.

Pros and cons of fossil fuels

Pros:
- The infrastructure is already in place
- They're cheaper in some cases
- They don't require as much land
- The sector employs millions around the globe
- They can be produced whatever the weather It's convenient to maintain the status quo

Cons:
- Their continued large-scale use is producing climate change that endangers humans, animals, and the Earth
- Air pollution resulting from coal kills 800,000 people per year
- Their locations disproportionately benefit the small number of countries where they are
- Jobs are falling, and are generally much less safe long-term than roles in the renewable energy sector
- Fossil fuel disasters are considerably worse – oil spills, oil fires, coal mining disasters, gas explosions, and mass poisonings can all kill or infect on a large scale
- They're finite, and will run out in the next century or so
- As the supply runs out, it may lead to economic, political, and social tensions within nations and between different countries
- The fossil fuel industry produced 37.1 billion tonnes of global CO_2 emissions in 2018

Information updated in July 2019.

Renewable energy vs fossil fuels

There really is no competition between the two categories of energy sources available to us. One is currently polluting the planet, and causing the climate to change in ways that will harm us and most species on Earth – while the other is clean, renewable, and infinite.

According to NASA, the ways in which we use fossil fuels as energy sources will lead directly to a variety of disastrous repercussions. For starters, we will have to deal with even more frequent wildfires, droughts and heat waves. By 2100, extremely hot days which previously happened once every 20 years will occur every two or three years.

Hurricanes will become stronger, longer, and generally more terrifying. A 2018 study published in scientific journal Nature predicted that climate change could increase rainfall totals for the worst hurricanes and cyclones by up to 30%. A different study in the American Meteorological Society journal also projected up to an 87% rise in the frequency of the most extreme hurricanes.

NASA expects sea levels to rise between one and four feet by 2100, and has pointed out that small islands like the Pacific nation of Tuvalu could be entirely submerged as a result. The UN has stated that a sea level rise of just 0.5m could displace a total of 1.2 million people from low-lying islands in the Caribbean Sea, Indian Ocean, and Pacific Ocean.

There's really no way to overstate the danger which climate change poses to these countries, or how much non-renewable energy sources contribute to the phenomenon. In 2017, Marshall Islands environment minister David Paul summed it up nicely when he told The Guardian: 'I can't stress enough that coal is by far the largest single barrier to staying within 1.5C of warming, and giving vulnerable countries like mine a chance of survival.'

And it's not just tiny islands that'll be affected. A 2017 UN report found that 40% of the world's population lives within 100km of a coastline. That's more than three billion people. Rising sea levels will affect everyone, forcing people on the coast inland, and potentially leading to overcrowding and resource scarcity.

That's not even factoring in the damage which rising sea levels could cause to harbours and ports, which the UN has estimated could be as high as £89 billion by 2050 and $293 billion by the end of the century.

In contrast, renewable energy sources are better for your health. They don't cause climate change, and generally have a minimal effect on the environment.

They've already taken over 22% of the world's electrical power production, and created 11 million jobs (despite many countries like the UK, France, and the Netherlands still giving more in subsidies to the fossil fuel industry than renewable energy producers). And best of all, they've managed this without melting any polar ice caps or increasing the frequency of hurricanes or wildfires.

Renewable energy sources also have the potential to act as a great equaliser in terms of energy production. Historically, a country has had to be blessed with oil, coal, and/or natural gas to be able to supply its own energy and sell it abroad. While a renewable future will still favour certain nations – namely those with abundant sunshine, wind, water, and geothermal energy – there will be more ways for nations to develop their energy industries.

Traditionally poorer countries like Mozambique, Tajikistan, and Kyrgyzstan have already made moves towards a future fully powered by renewable energy. This could mean being able to avoid an overreliance on getting energy from global forces like Saudi Arabia, China, and Russia – and could even lead them to develop their economies by exporting renewable energy.

For instance, as of 2016, Ethiopia derived 93% of its energy from renewable sources – and a new, multi-billion-pound geothermal project may provide further opportunities to boost the nation's economy.

As well as being easier for different countries to produce, renewable energy is also becoming cheaper than fossil fuels. According to a 2018 International Renewable Energy Agency report, out of all the global energy production centres due to be commissioned in 2020, 'over three-quarters of the onshore wind capacity and four-fifths of the solar PV project capacity… should produce cheaper electricity than any coal, oil, or natural gas option.'

Summary

We've reached the tipping point.

It's been known for years that fossil fuels are hurting the planet we live on, endangering animals and humans alike. But now, finally, renewable energy makes sense on a financial as well as a moral level.

And it's not just hypothetically possible to power whole countries with renewable energy; it's already been done. Albania, Lesotho, Nepal, and Paraguay have proved this fact, while nations like Iceland and Costa Rica have also come close to 100%.

Renewable energy sources are less catastrophic than fossil fuels, both in their effect on the climate and in terms of how few disasters – like oil spills and gas fires – they produce.

But despite all of these facts, it's proving difficult to convince countries to use more renewable energy when they're trying to develop their economies as quickly as possible. 2018 saw a 2% rise in CO_2 emissions – the largest increase in seven years, according to oil and gas company BP, and one which took CO_2 levels to a record high.

And some of the main offenders are the nations with the biggest economies in the world. For example, the US accounted for almost all of the 2.4% global increase in oil production, and nearly half of the rise in global natural gas production.

It's up to you to decide whether you want to be part of the renewable revolution, or would rather turn a blind eye.

It may also help you to sell your home one day, as 65% told our latest National Home Energy Survey that they were likely to buy a house with solar panels.

Last updated 24th May 2022

The above information is reprinted with kind permission from the eco experts.
© 2022 the eco experts

www.theecoexperts.co.uk

Top 5 fastest-growing renewable energy sources around the world

Renewable energy sources are an important tool in divesting from fossil fuels, reducing emissions and fighting climate change. Major economies, like the UK, have pledged to cut their emissions to net-zero by mid-century, or like China, 2060. By 2035, renewable energy sources are going to account for more than half of global electricity production. But what will be powering this eco-friendly future? Here are five of the fastest-growing renewable energy sources across the world and how they are being used.

Hydropower

First on our list of the fastest-growing renewable energy sources, hydropower is the most widely used form of renewable energy in the world, producing 1 295 gigawatts of energy. This amounts to 54% of the global renewable power generation capacity. The most common hydropower comes from water in dams. The water gets released from the reservoir to drive the turbines which generate the power. However, it can also be used by the natural running of a river or tide to drive the turbines. Because hydropower can be generated quickly, it is used as a pumped-storage plant that can provide backup energy at short notice.

China has the biggest hydroelectric generation in the world. The Three Gorges generate 22.5 GW. China has made large investments in hydropower so that they are not as reliant on coal. It gets around 15% of its energy from hydropower. In Southwest China many of its rivers are high above sea level, and China knows that hydropower is the only renewable energy they can use on a big scale, therefore it makes big investments in the sector.

Wind energy

Wind turbines are normally used to get kinetic energy from the wind to generate energy. Wind energy is the second most used renewable energy source in the world, producing 563 GW, and produces 24% of the world's total renewable energy generation capacity.

The UK is the sixth-biggest producer in the world, producing 13 603 megawatts. Offshore wind farms are constructed in bodies of water, and in the UK, wind farms power the equivalent of 4.5 million homes. Onshore wind farms deliver less energy for the UK, only providing around 10% of UK energy by 2020, despite being the most cost-effective alternative for new electricity in the UK compared to traditional fossil fuels.

Solar power

Solar power works by converting light from the sun into energy. The UK is installing solar panels faster than any other European country. Solar power has had an annual average growth rate of 25% over the last five years across the world. Spain is a major producer of solar power, contributing 75% of the global concentrated solar power.

The biggest solar power plant in the world is located in the United Arab Emirates. The Noor Abu Dhabi solar project produces 1.17 GW, producing enough electricity to power the demand of 90 000 people. It will reduce its carbon footprint by 1 million metric tonnes a year, which is the same as taking 200 000 cars off the road, proving the UAE is making good progress in the fight against climate change even though the country is still oil-dependent.

Bio-power

Modern biomass includes biofuels and wood pellets as well as traditional ones that were already used, such as agricultural by-products. These products are then burned to create steam, which powers a turbine that generates energy. China, the UK, and India accounted for more than half of the world's total bioenergy capacity expansion in 2018, while bioenergy provides 11% of the UK's energy and bioenergy.

Geothermal

Last on our list of the fastest-growing renewable energy sources, geothermal energy is thermal energy generated and stored in the earth. Globally, geothermal production exceeded 13.2 GW in 2018. One-third of green energy that is made using geothermal sources is electricity. Iceland is one of the world's biggest producers of geothermal electricity, producing 26.5% of the country's electricity and 87% of their housing and building needs from natural hot water sourced underground. In the UK, geothermal energy is not the most viable option as the ground is not hot enough, but it has increasingly been using shallow resources from the upper crust of the earth that is heated by the sun. With the help of ground source heat pumps, the energy is extracted.

While these fastest-growing renewable sources of energy are encouraging, there is still much work to be done in the realm of renewable energy. It is vital that governments around the world shift to using renewable energy to power their countries to give us the best shot at meeting the targets under the Paris Agreement.

10 March 2021

The above information is reprinted with kind permission from EARTH.ORG
© 2022 Earth.Org

www.earth.org

Hornsea 2, the world's largest windfarm, enters full operation

Ørsted is proud to announce that the world's largest installed windfarm, Hornsea 2, is now fully operational.

The 1.3GW project comprises 165 wind turbines, located 89km off the Yorkshire Coast, which will help power over 1.4 million UK homes with low-cost, clean and secure renewable energy. It is situated alongside its sister project Hornsea 1, which together can power 2.5 million homes and make a significant contribution to the UK Government's ambition of having 50 GW offshore wind in operation by 2030.

The Hornsea Zone, an area of the North Sea covering more than 2,000 sq km, is also set to include Hornsea 3. The 2.8GW project is planned to follow Hornsea 2 having been awarded a contract for difference from the UK government earlier this year.

Hornsea 2 has played a key role in the ongoing development of a larger and sustainably competitive UK supply chain to support the next phase of the UK's offshore wind success story. In the past five years alone, Ørsted has placed major contracts with nearly 200 UK suppliers with £4.5 billion invested to date and a further £8.6 billion expected to be invested over the next decade.

Ørsted now has 13 operational offshore wind farms in the UK, providing 6.2GW of renewable electricity for the UK – enough to power more than 7 million homes. Hornsea 2 makes a significant contribution to Ørsted's global ambition of installing 30 GW offshore wind by 2030. Ørsted currently has approx. 8.9 GW offshore wind in operation, approx. 2.2 GW under construction, and another almost 11 GW of awarded capacity under development including Hornsea 3.

Patrick Harnett, Vice President UK Programme, Ørsted, said: 'This project has been an amazing endeavour. To build the world's largest offshore windfarm during a global pandemic has been a challenge that the team have overcome with flying colours. I am so proud of how our team has worked together to safely deliver this remarkable project. A huge thank you to all those involved in making it happen.'

Darren Ramshaw, Vice President, and Head of UK East Coast Region, Ørsted said: 'Thanks to all those who have built the windfarm, our operations and maintenance teams are already at work. Hornsea 2 is another brilliant addition to our East Coast portfolio. Now Hornsea 2 is fully operational, our total capacity from Grimsby is 3.8GW, providing enough electricity to power 3.3 million homes. And it does not just stop there, by 2030, we will be on track to power over a quarter of UK households. Our teams here work tirelessly on making sure Britain is powered by clean energy.'

Duncan Clark, Head of Region UK at Ørsted, says: 'The UK is truly a world leader in offshore wind and the completion of Hornsea 2 is a tremendous milestone for the offshore wind industry, not just in the UK but globally. Current global events highlight more than ever the importance of landmark renewable energy projects like Hornsea 2, helping the UK increase the security and resilience of our energy supply and drive down costs for consumers by reducing our dependence on expensive fossil fuels.

'Not only will Hornsea 2 provide low cost, clean energy for millions of homes in the UK, it has also delivered thousands of high-quality jobs and billions of pounds of investment in the UK's offshore wind supply chain. We look forward to working with government and industry colleagues to continue to accelerate the deployment of offshore wind for the benefit of homes and businesses across the country.'

Hornsea 2 has also seen a number of other significant milestones during construction including the award winning Thrive safety programme, a legacy project for the wind farm and designed to be an innovative learning centre that will continue to benefit industries across the Humber.

The project has also supported innovation, such as the development of the 'Get Up Safe' (GUS) motion-compensated lifting system by Pict Offshore. The company, which following development work with Ørsted received support from Scottish Enterprise, developed a new technology for lifting technicians onto wind turbine platforms that is safer, simpler and more effective.

In October 2020, the project donated £1 million to the Horizon Youth Zone that will be built in the centre of Grimsby. This hub will be a brilliant place for young people to meet and learn new skills.

Just over a year ago, Ørsted announced its 1000th turbine in UK water and welcomed a state-of-the-art new service operations vessel to their fleet, Wind of Hope.

Quick facts about Hornsea 2 Offshore Wind Farm

- 165 turbines delivering 1.3GW of renewable electricity
- The wind farm spans an area of 462 sq. km – equal to more than 64,000 football pitches or 31 Lake Windermeres
- Each turbine blade is 81m long and the blade tip reaches more than 200m above sea level
- One revolution of the turbine blades can power an average UK home for 24 hours
- 390km of subsea export cables take the power generated from Hornsea 2 to the shore at Horseshoe Point in Lincolnshire

31 August 2022

The above information is reprinted with kind permission from Ørsted UK.
© 2022 Ørsted

www.hornseaprojects.co.uk

How do UK energy prices compare with the rest of Europe?

Energy prices in the UK just keep on going up… and up.

By Elizabeth Atkin

Following news that October will see Ofgem's price cap soar to £3,549 for an average annual bill, and forecasts of an even higher hike from January 2023, many are really feeling the cost of living pinch.

You might be wondering what the energy price situation is like over in Europe, in countries such as France and Germany – as you mull over the available help or make plans to cut your usage.

Here is how the UK compares with the rest of Europe.

What are energy prices like elsewhere in Europe?

The UK is paying the second highest amount of electricity in Europe, only topped by the Czech Republic, according to the Household Energy Price Index (HEPI).

It can be difficult to do a direct comparison of energy prices across countries, so HEPI looked at household electricity and gas costs in European capital cities in July 2022, focusing on the average price per kWh.

HEPI uses an artificial currency called purchasing power standards (PPS) to create a more accurate comparison.

The latest figures in July show the UK is paying more than 51.85 pps per kWh of electricity, just under Czech Republic's 52.15 pps. To compare, France is paying 23.2 pps for kWh, and Germany 35.93 pps.

The data also shows the cheapest countries for electricity are Malta (13.95 pps), Switzerland (13.8 pps), and Norway (12.83 pps).

Which European country has the highest electricity prices?
According to HEPI, the top ten most expensive European countries for electricity are:

1. Czech Republic
2. UK
3. Italy
4. Estonia
5. Denmark
6. Latvia
7. Cyprus
8. The Netherlands
9. Belgium
10. Germany

Which European country has the highest gas prices?
According to HEPI, the top ten most expensive European countries for gas are:

1. Bulgaria
2. The Netherlands
3. Greece
4. Latvia
5. Czech Republic
6. Sweden
7. Estonia
8. Spain
9. Denmark
10. Portugal

As for gas, it's a slightly different picture. Using the same parameters, the UK falls in the middle of the table (at 15.21 pps) – and is behind table-topper Bulgaria (24.82 pps), The Netherlands (24.54 pps), Czech Republic (20.12 pps), and Germany (15.51 pps).

However, the UK's gas costs are still above the European average they've calculated and come in more expensive than Belgium (36.13 pps), France (23.2 pps), or Ireland (28.26 pps).

Hungary – which in late August 2022 signed a new deal for gas with Russia's Gazprom provider – was the very cheapest in July (at 4.32 pps), with Serbia (5.99 pps), and Slovakia (6.52 pps) close behind.

Why are UK energy prices so high right now?

The cost of energy has been rising all over Europe, but why is it so high in the UK?

Wholesale prices for natural gas have been rising since last year – pushing dozens of UK energy companies out of business (though said prices did experience a drop towards the end of August 2022).

The upped cost is due to a post-pandemic rise in demand for energy, as well as the war in Ukraine – as Russia is one of the world's biggest gas providers – which has created instability in the energy market.

In the UK, we don't rely on Russian gas as much as other European countries do (just over 3% to their 40%), though we do rely on gas a lot.

We also have few places to store it, meaning the UK buys more in the short term – and we also have less nuclear and renewable energy sources than other parts of Europe, MarketWatch explains.

Robert Buckley, Head of Relationship Development at Cornwall Insight, previously told Metro.co.uk:

'One of the most significant contributors to energy market volatility is the war in Ukraine, for which there is of course no timetable for a resolution.'

He added that it's 'difficult' to predict when energy prices might drop, noting: 'Our price predictions [for] later in the decade have gone down slightly, but they remain above pre-2021 levels until 2030 and likely beyond.

'The increase in renewables deployment in Europe due to the conflict in Ukraine has helped push our price predictions down slightly.'

1 September 2022

The above information is reprinted with kind permission from Metro & DMG Media Licensing.
© 2022 Associated Newspapers Limited

www.metro.co.uk

How much of the UK's energy is renewable?

With the UK aiming to reach net zero by 2050, a crucial part of the strategy is to transition to an electricity system with 100% zero-carbon generation and much of this is expected to come from renewable energy.

Renewable energy is already part of our electricity mix (the different energy sources that make up our electricity supply), but how much are we using currently, and how much more will we need in order to reach net zero?

Why is renewable energy important?

Clean power generation is front-and-centre of the UK's strategy to reach net zero by 2050, with the government setting energy providers a target for all electricity to come from 100% zero-carbon generation by 2035.

Burning fossil fuels to create electricity has long been a major contributor in the emission of greenhouse gases (GHGs) into our atmosphere. As renewable energy sources emit low or no carbon emissions, they are considered vital in the race to tackle climate change.

What renewables are used to generate electricity?

Today, there are four main renewable energy sources used to power the UK: wind, solar, hydroelectric and bioenergy. They harness the natural power of the sun, our weather, our waterways and tides, and organic materials to generate electricity.

Currently, the majority of the electricity entering the national grid from a single energy source is natural gas. Natural gas is a largely imported fossil fuel and can emit harmful GHGs such as carbon dioxide (CO_2) when burned to generate electricity.

How much of our energy currently comes from renewable sources?

Today, renewable energy sources make up a significant proportion of the electricity mix that powers our homes and businesses. And the UK is well on its way to creating an electricity system that's wholly based on renewable and carbon-free sources.

2020 marked the first year in the UK's history that electricity came predominantly from renewable energy, with 43% of our power coming from a mix of wind, solar, bioenergy and hydroelectric sources.

The UK is on the cusp of producing its trillionth kilowatt hour (kWh) of renewable energy since 1970. While it took 47 years (from 1970 to 2017) to produce the first half trillion, we will have produced the second half trillion between 2017 and 2023 alone.

How has our use of renewables changed?

By the end of 1991, renewables accounted for just 2% of all electrical generation in the UK. By 2013 this figure had risen to 14.6%.

2017 placed Britain into the position as one of Europe's leaders in the growth of renewable energy generation. Only countries like Iceland, Norway and Sweden, who had more established renewable schemes, used more on a relative scale.

In 2019, zero-carbon electricity production overtook fossil fuels for the first time, while on 17 August renewable generation hit the highest share ever at 85.1% (wind 39%, solar 25%, nuclear 20% and hydro 1%).

In the last quarter of 2021, individual renewables contributed the following:

- **Wind power** contributed 26.1% of the UK's total electricity generation in Q4 2021, with onshore and offshore wind contributing 12% and 14% respectively.
- **Bioenergy,** the burning of renewable organic materials, contributed 12.7% to the renewable mix.
- **Solar power** contributed 1.8% to the renewable mix – this represented a 24% increase compared to Q4 2020, due to a 0.7 gigawatt (GW) increase in installed capacity.
- **Hydropower,** including tidal, contributed 2.1% to the renewable mix.

Breaking records: The UK's renewable energy in numbers*

2020 was the UK's highest year on record for renewable generation so far, and we've been breaking records for renewables ever since.

- Zero-carbon power in Britain's electricity mix has grown from less than 20% in 2010 to nearly 50% in 2021. In contrast, power provided from fossil fuels was down to roughly 35% in 2021 compared with over 75% in 2010.
- In 2020 renewables accounted for more than 43.1% of the UK's total electricity generated, at 312 terawatt hours (TWh). This outstripped fossil fuels over the course of a year, for the first time in the nation's history.
- 2020 also saw UK have its longest run of coal-free power, with a total of 68 days between 10 April and 16 June. This is the longest coal-free period since the industrial revolution, which began in the mid-1700s!
- Zero-carbon generation overtook fossil fuel consumption in 11 months of the year in 2021.
- 2021 was the second highest year for renewable generation on record, after 2020.
- On 5 April 2021, the UK achieved its lowest ever carbon intensity at 39 grams of CO_2 per kWh, due to reduced use of fossil fuels for electricity generation. This was made possible by a 60% increase in the rate of renewable capacity installed in 2021 (compared to 2020).
- 25 May 2022 holds the record for the maximum amount of wind power generation, at 19.9 GW.

How long will it take to switch to renewable energy?

It's important to remember that the aim is not for renewables to be our sole provider of energy, but they will play a major part in the energy mix alongside other clean and green energy sources.

This said, renewable energy has grown ten-fold since 2004 and the UK looks on track to continue to increase renewable generation – with renewable energy sources making up 42.8% of the UK's total electricity generation between October and December 2021.

It's anticipated that the UK's renewable capacity will increase dramatically over the next decade. Plans are already in action to increase offshore wind's output from 11 GW to 50 GW by 2030 – helped by a £200 million government cash injection and financial incentives. Meanwhile, solar capacity could grow five-fold from 14 GW to roughly 70 GW in the same period.

Combine renewables with other low-carbon electricity sources, such as nuclear (16%), and it indicates that our green infrastructure is heading in the right direction to be capable of reaching the government's 2035 target; and ultimately reaching net zero in the specified time frames.

2022

* Numbers quoted here as the current record for wind power etc. may of course be exceeded in the future.

The above information is reprinted with kind permission from National Grid.
© National Grid 2022

www.nationalgrid.com

Why the UK's unfair energy market is unlikely to spearhead a green transition

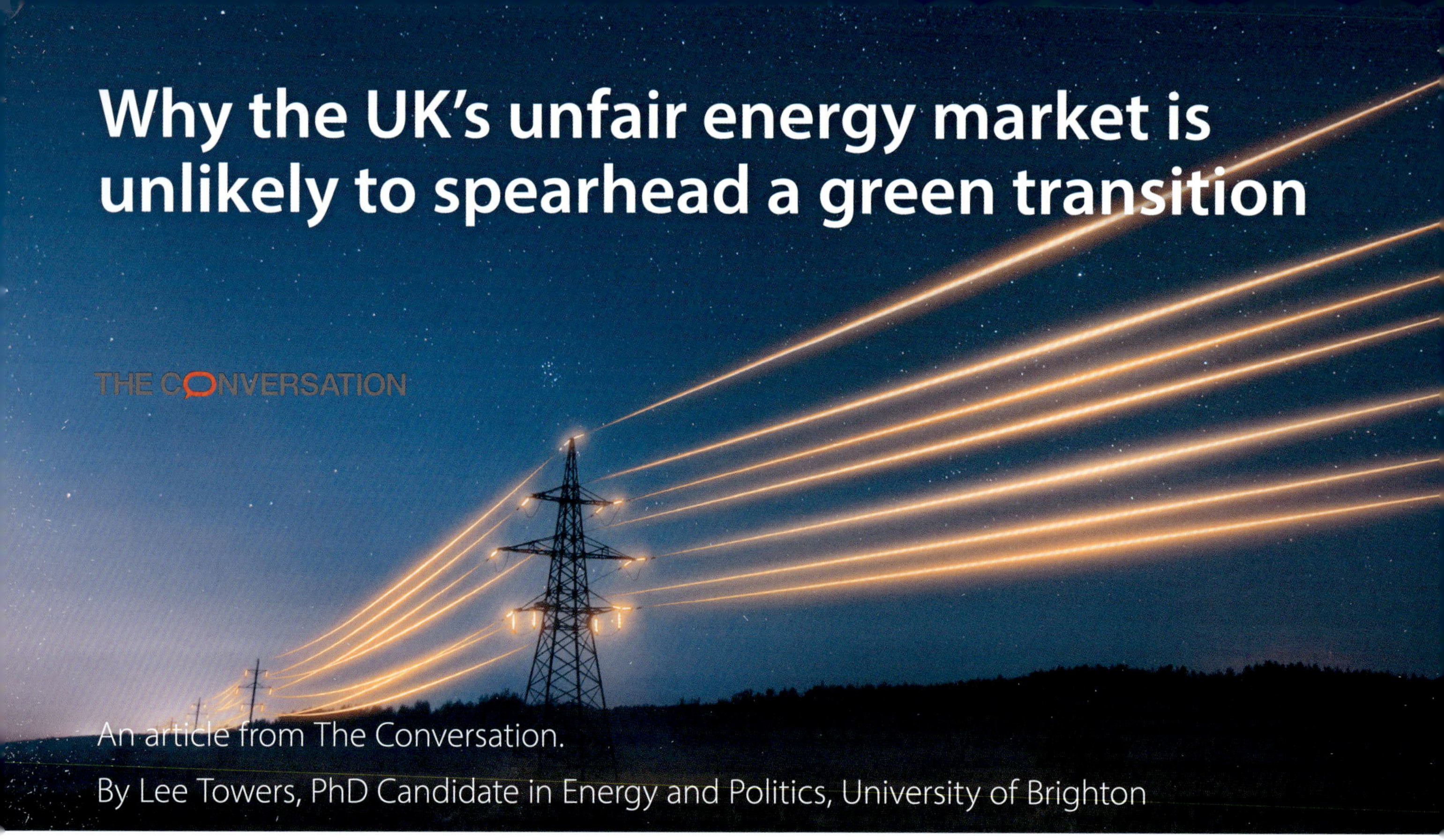

An article from The Conversation.

By Lee Towers, PhD Candidate in Energy and Politics, University of Brighton

Millions of people will pay more for gas and electricity in the UK by the autumn of 2021 as the energy regulator, Ofgem, has removed a cap on prices. For some households, this could mean bills rising by as much as £153 (US$212) a year, potentially pushing an extra 500,000 homes into fuel poverty.

Those heftier energy bills also fund much of the UK government's flagship policies for the low-carbon transition. On top of their energy use, every home in the country is paying extra on their bill to cover the cost of retrofitting programmes to increase the energy efficiency of homes, help for those in fuel poverty and subsidies for renewable

Who foots the bill for decarbonisation?

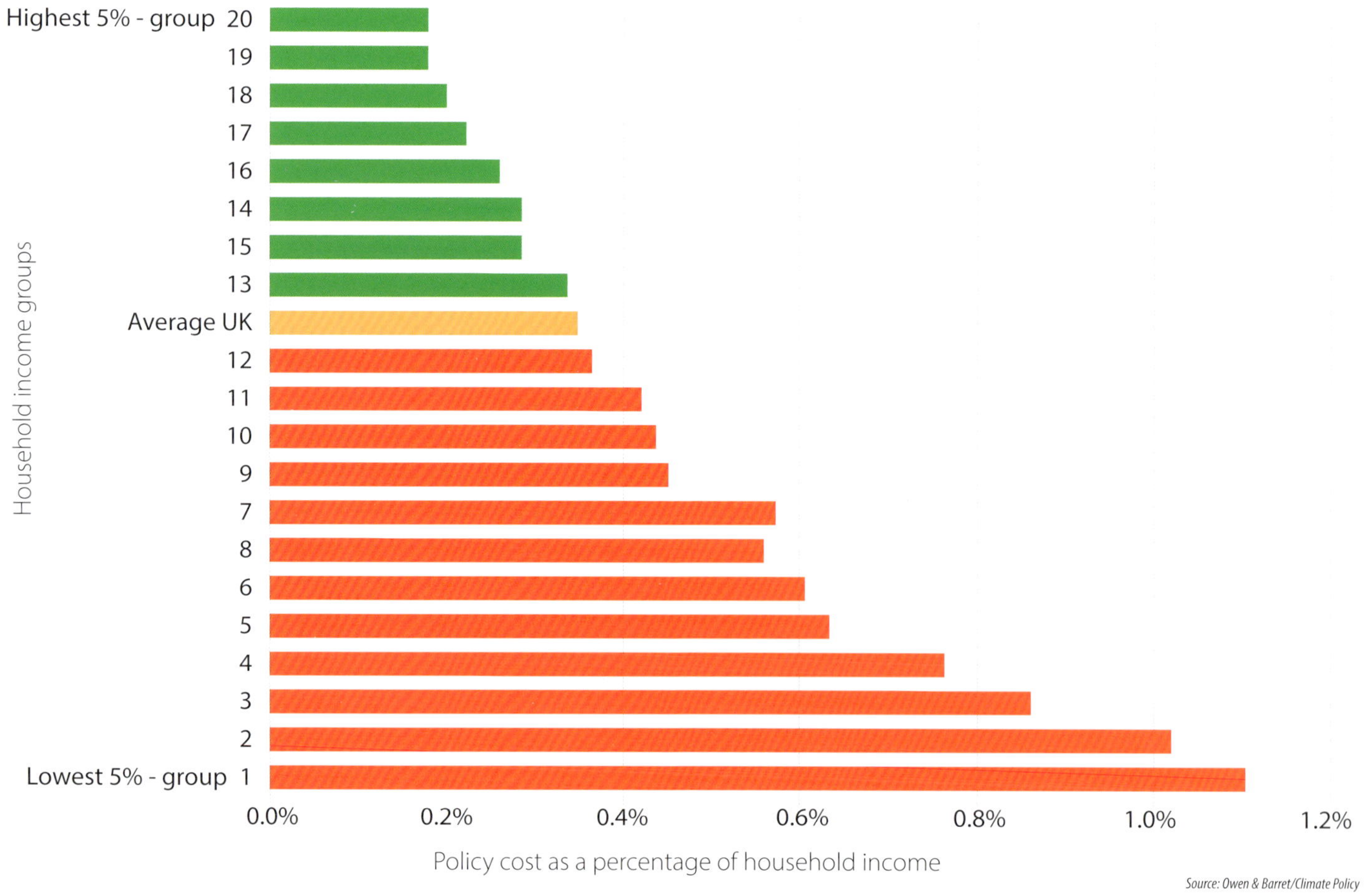

Source: Owen & Barret/Climate Policy

Higher income groups use significantly more energy than those on lower incomes.

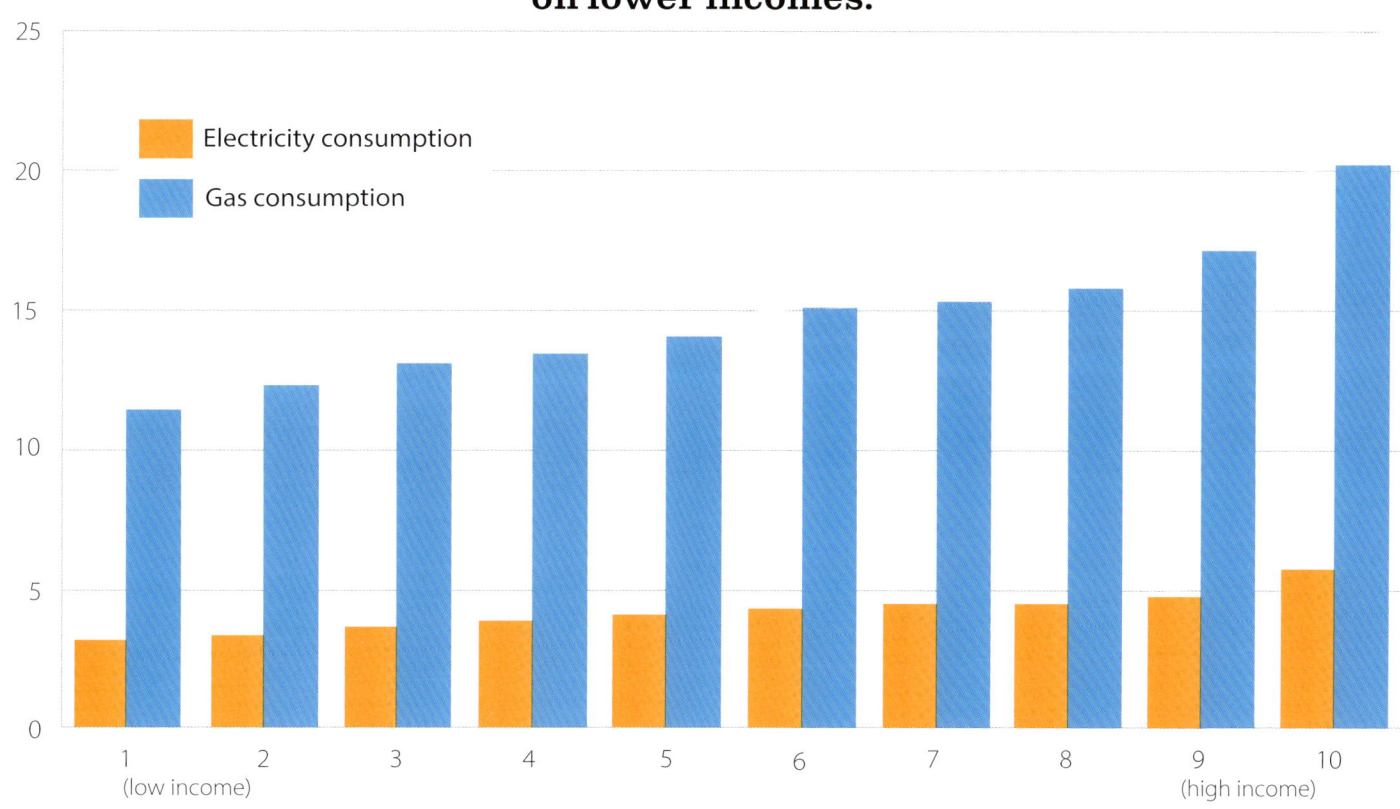

Source: Garman & Aldridge/IPPR

generation. All of these costs are added to energy bills at a flat rate.

According to a 2015 report, this means, in practice, that those on the lowest incomes pay a six times higher share of their income for the transition than the highest income group, who also happen to have the highest CO_2 emissions on average.

Through energy bills, people in the lowest income groups effectively self-fund their own fuel poverty support, including measures like the warm home discount – a one-off winter payment of £140 towards energy bills – while also paying towards measures that mainly benefit higher income groups, like subsidies for rooftop solar panels.

Academic Brenda Boardman warned about this problem in the 1990s. Not only is this not a fair way to fund the national effort to decarbonise the UK's energy system, but that same unfairness is slowing the speed and reducing the motivation for a transition in the first place.

The big fix

The UK's energy market is dominated by six multinational corporations. On the website of Ofgem, the energy regulator, a page documents the multiple infractions of these companies, known as the Big Six, from their treatment of vulnerable customers to their failure to fulfil obligations to reduce the carbon intensity of their gas and electricity.

Ofgem's answer is a voluntary redress scheme which companies under investigation pay into. This often funds programmes which advise vulnerable customers, such as the chronically ill, on how to navigate an (often intentionally) bewildering energy market. Npower paid a record fine in 2015 of £26 million. The total number of fines and redress payments made in 2020 reached £71.3 million.

Had this money been spent directly on low-cost renewable generation, 57 megawatts of wind energy could have been installed, enough to power around 40,000 homes annually.

These six companies made profits of nearly £2 billion in 2015 from their standard tariffs. People are often placed on these automatically and tend to remain on them, with only the most nimble customers (around 30% in 2016) switching. The government noted that those least likely to switch to cheaper tariffs earn less than £18,000 a year, are aged 65 and over, with a disability, or live in social housing or the private rented sector. As a result, a significant proportion of these profits are extracted from those least able to pay.

In fact, many people living in private rentals and social housing (20% and 17% of the population respectively) are effectively excluded from the choice of installing the solar panels and electric vehicle charging points their energy bills finance, because such freedoms tend to depend on home ownership.

UK consumers also pay some of the highest pre-tax rates for electricity in Europe, and energy costs in general are higher than they should be considering falling gas prices since 2014 and more efficient boilers and smart meters. Meanwhile, the third of the bill which pays for the maintenance and environmental upgrade of energy infrastructure is taken at the same rate from billionaires as it is from those on Universal Credit. People without solar panels will continue paying for the essential network changes that incorporate the growing amount of renewable energy, and those who cannot afford

Electricity prices from July to December 2015, including tax (p/kWh).

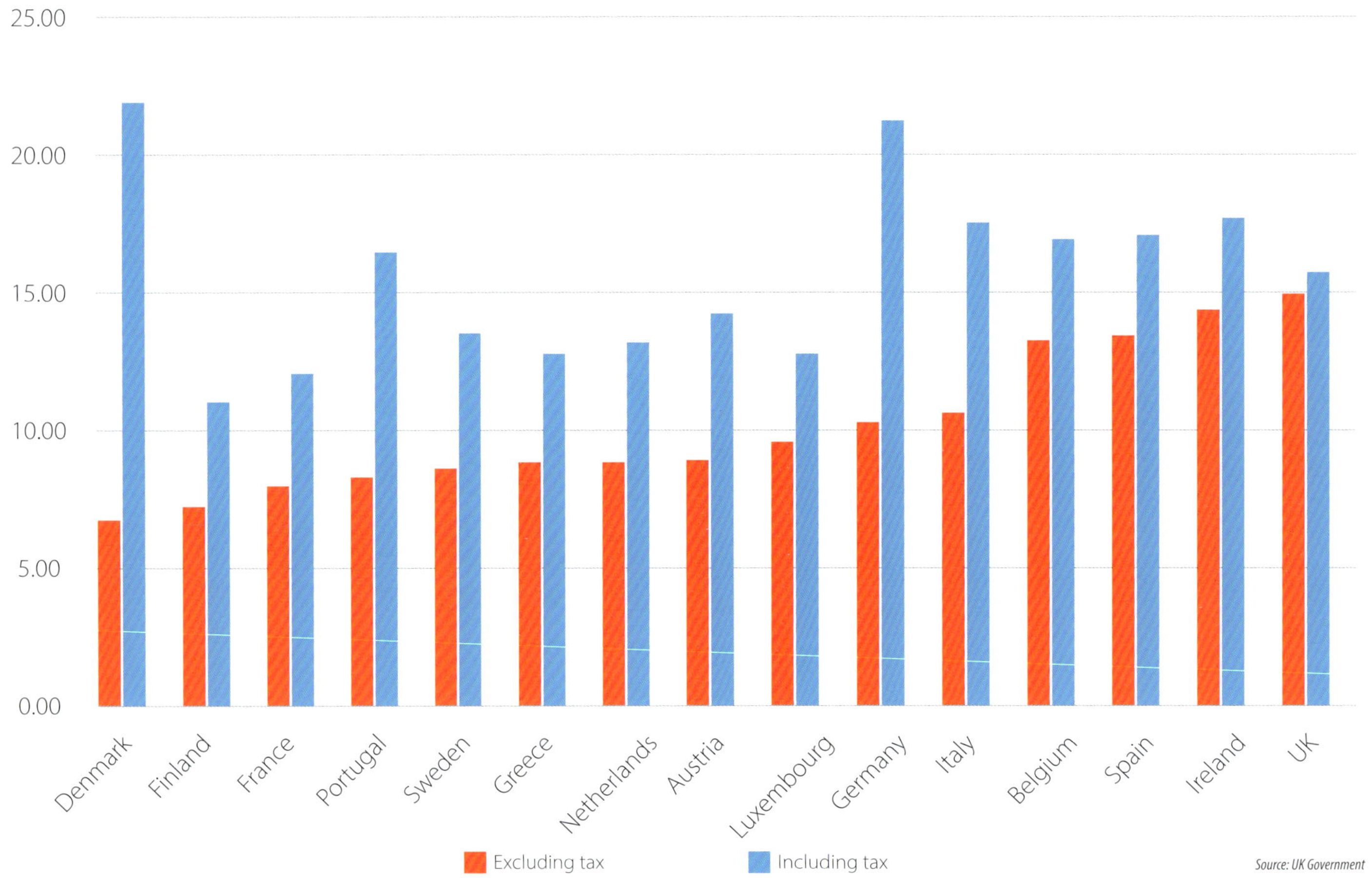

[Bar chart showing electricity prices excluding tax (orange) and including tax (blue) for: Denmark, Finland, France, Portugal, Sweden, Greece, Netherlands, Austria, Luxembourg, Germany, Italy, Belgium, Spain, Ireland, UK]

Source: UK Government

to swap their fossil fuel burning car for an electric vehicle could end up having to travel further and emit more to get fuel, as retailers cease supplying or close their facilities.

While consumers fund measures which cut emissions from the UK's energy system through bills, government policies effectively subsidise fossil fuels, mostly through forgone tax revenue. These UK subsidies result in effective investment in fossil fuel production of €11.6 billion a year, compared to the €7.76 billion invested in renewables. In this situation, the renewables consumers fund are more likely to add to the energy generated by fossil fuels, rather than replace it.

At the same time, renewable subsidies like the feed-in-tariff, which paid people for the excess energy they generated with solar panels, have been axed. This makes it harder for people to pay to install solar power at home. Clearly, the transition to low-carbon energy could be quicker if government policy didn't work against the best intentions of the public.

And the public are aware. One study in 2019 found widespread misgivings about excessive profits, a lack of transparency and close ties between the government and big energy companies. If people don't trust the institutions tasked with overseeing the end of the fossil fuel era, how will they be persuaded to make the necessary changes to their own lives?

5 August 2021

The above information is reprinted with kind permission from The Conversation.
© 2010-2022, The Conversation Trust (UK) Limited

www.theconversation.com

Rising sales of electric vehicles show they are here to stay

More needs to be done so electric vehicles can reach their potential to reduce carbon emissions. But surging sales show they are here to stay.

By Jeremy Lovell

Electric cars are flying off the world's shelves like there is no tomorrow. Most auto manufacturers now offer at least one full-battery model and a hybrid version, and some have even declared their aim of going fully electric within a decade or two.

Far from suffering along with the rest of the world's economies due to COVID-19 effectively closing down normal world commerce, sales of electric vehicles (EVs) not only bucked the trend but accelerated through it, surging more than 40% year-on-year to a record three million new registrations in 2020, while those of internal combustion engined (ICE) models fell 6%.

There are a multitude of reasons for the boom in EV and hybrid sales. They include government subsidies, legislation governing new ICE vehicle sales and emissions limits, rising EV ranges, expanding charging infrastructure and gradual cost reductions.

But there can be little doubt the pandemic not only provided hard evidence of the major improvement in urban air quality and slump in global carbon emissions during the various lockdowns and travel bans. It also underscored the interconnectedness and mutual responsibilities of nations and communities within them.

Before the pandemic, climate change may have been more of a concept than a reality to many despite the sharp increases in floods, fires and droughts, and the carbon reductions of the 2008-09 world recession may have been a distant and fading memory. But the unprecedented impacts of the pandemic showing the world as a sum of its parts rather than a series of disconnected entities brought it sharply back into focus.

Declarations by various governments that they intend to 'build back greener' from their decimated economies will have added fuel to the fire.

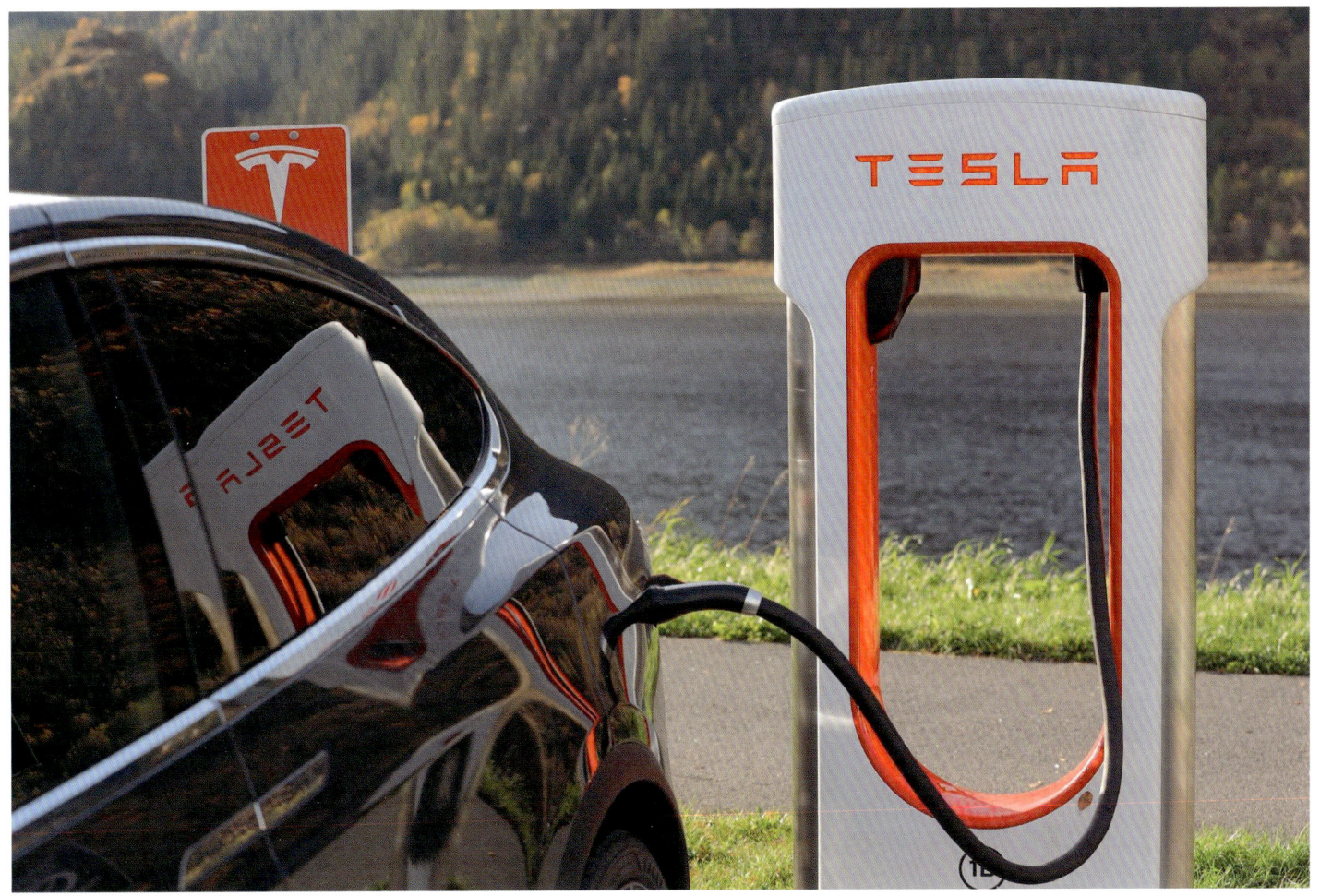

Exponential growth in sales of electric vehicles

Carbon emissions from transport vary enormously from country to country and region to region, but are broadly credited with producing over one quarter of the global total. So they are a prime target for action, whether through electrification, biofuels, engine efficiencies, alternative fuels such as hydrogen, travel reductions or simply getting off four wheels and on two.

The latest figures show there are now more than 10 million electric cars on the road globally, along with about one million electric vans, heavy trucks and buses.

This is a fraction of the hundreds of millions of gasoline and diesel-engine vehicles already on the roads and likely to remain there for decades, regardless of any looming bans on sales of new ICE models. But the growth is exponential.

A report last month from the International Energy Agency — Global Electric Vehicle Outlook 2021 — showed sales of electric vehicles in the first three months of the year up 2.5 times from the same, mostly pre-COVID period of 2020.

It also predicted the number of electric cars, vans, trucks and buses on the roads globally would erupt to between 145 million and 230 million by 2030, depending on the levels of climate ambition both declared and actually achieved by the more than 190 signatory nations to the 2015 Paris Climate Agreement.

There is no question that electrification facilitates the advent of automated or driverless cars. But what impact they might have on vehicle numbers and emissions and how long they will take to gain general acceptance is open to question — although initial acceptability is likely to be higher among younger people.

Automakers had collectively some 370 electric car models on offer globally last year. Companies making 18 of the top 20 models, accounting for 90% of global auto sales, have said they intend to both add more EV models and expand into electric-powered light freight vehicles, with road freight a major target market for electrification

The future is clearly visible.

According to the IEA report, the swap-out to electric transport could cut greenhouse gas emissions by between one third and two thirds in 2030 compared to business as usual, depending on whether countries meet or raise their climate ambitions.

But that is only part of the story.

A wholesale switch to electric vehicles would by its nature slash carbon emissions from transport. But the picture is far less clear if more fossil fuels are burned to generate the extra electricity needed to recharge the batteries. And that leaves well to one side questions over the rare earths needed to make the batteries, the carbon emitted to mine and refine them, and what use is made of the power packs once their efficiency has been depleted over time.

Three questions to consider:

1. How has the COVID-19 pandemic made the reality of climate change more apparent to people?

2. In what ways can electric vehicles contribute to carbon emissions?

3. Would you consider buying an electric vehicle?

Similar questions hover over the development and expansion of hydrogen fuel cells, which can power vehicles while emitting water. Such cells are an alternative to electricity, but currently they are uncommercial except in countries like Iceland which have vast quantities of clean, geo-thermal power to produce the hydrogen.

Leading automakers agree that while they have firm goals to go electric as soon as possible, the rate of change will vary hugely depending on countries' access to clean electricity.

Germany's Volkswagen, whose public image was shattered in 2015 when it was exposed as having deliberately falsified emissions data from its diesel-powered vehicles, has vowed to go 'climate-neutral' and said it will take up to 'two car model lifecycles' to achieve that goal.

But automakers accept that the future is clearly visible. While events and circumstances may accelerate or impede progress at different times in different countries, the momentum makes a conversion to EVs unstoppable, as does its target audience — the young who will have to live with the consequences of the climate change that is already in the pipeline.

As Tom Burke, chairman of influential environmental think tank E3G, summed up: 'The young don't need to be sold the idea. They already get it. They can see what their parents have done to the planet, and they want none of it.'

Jeremy Lovell was a correspondent for Reuters for more than 23 years in Europe, Asia and Africa. He covered Dutch, Belgian, British and South African elections, the EU's Exchange Rate Mechanism crisis, Belgian pedophile murders, NATO going to war for the first time, Zimbabwean farm invasions and climate change, energy and the environment.

10 May 2021

The above information is reprinted with kind permission from News Decoder.
© News Decoder 2022

www.news-decoder.com

Electric cars are greener than petrol cars – but they're far from perfect

Switching to electric cars is essential, but it's not enough. Our transport system needs a rethink.

Are electric cars really greener than petrol or diesel cars? If you've ever felt confused about this question, you're not alone. Stories about the (very real) problems with electric cars are sometimes used to argue in favour of the fossil-fuelled status quo. But the reality is that an electric car has about half the climate impact over its lifetime compared to an average EU car today.

In fact, rapidly switching from fossil fuelled cars and vans to electric vehicles is one of the most important things the government can do for the climate.

That's because driving makes up a huge share of the UK's carbon footprint. Right now, nearly four out of every five miles travelled in the UK happens in a car. For the past 60 years, we've been building our towns, cities and entire lives around widespread car ownership. And that won't change overnight.

It's important to reduce the need for cars by giving people a real alternative, or simply reducing the need to travel in the first place.

But to cut carbon fast enough to avoid the worst of climate change, we're going to have to replace at least some fossil-fuelled cars with electric ones. Let's take a look at how they compare.

Electric cars vs petrol cars

There are two main reasons why electric cars are so much better for the environment than petrol and diesel.

1. Electricity is getting cleaner all the time

While conventional cars will always need dirty fossil fuels, electric vehicles can (and increasingly do) run on renewable energy. In the UK, the carbon footprint of electricity is falling fast, and Greenpeace is campaigning for an 80 percent renewable grid by 2030. Every year this trend continues, electric cars increase their advantage.

As the electricity grid gets cleaner, the carbon impact of manufacturing falls for all new cars. But once they're actually on the road, powering a petrol car is as polluting as ever.

2. Electric motors are much more efficient than conventional engines

Despite more than 100 years of refinements, the internal combustion engine used in cars just isn't that good at converting fuel into movement. Even in the most efficient petrol engines, only around 12-30 percent of the energy in the fuel ever makes it to the wheels or other useful functions. The rest is wasted as noise and heat.

Electric motors, by contrast, are more like 77 percent efficient – they get more than twice as many miles out of the same amount of energy. This efficiency gap is so big that even in Poland where most electricity comes from coal-fired power stations, electric cars emit about 25% less carbon than their fossil fuelled equivalents.

How to make electric cars better

While electric cars are undoubtedly less harmful than petrol cars, frankly that's a pretty low bar. It's important to understand the problems with today's electric cars, and demand that the industry do better.

Air pollution

There's been lots of debate about cars causing air pollution in cities. But most people don't know that some of this pollution comes from cars' tyres and brakes – not just the exhaust pipe. All that weight and friction scrapes off tiny particles of plastic and other nasty stuff, which ends up in people's lungs, or washing into rivers. So when it comes to pollution, going electric only solves part of the problem.

Mining

There are also huge problems with some of the materials that make up today's electric vehicle batteries. As production ramps up, these urgently need to be fixed. Cobalt production is linked to child labour in the Democratic Republic of the Congo. Some Indigenous communities are resisting lithium mining on their land in South America.

We could pit these outrages against the horrors of fossil fuel extraction, but it doesn't make them any less ugly. That's why the transition away from fossil fuels can't just be about swapping one set of machines for another. We need to avoid using more than we need, and clean up the industries that supply it to us.

While it's impossible to burn oil ethically and sustainably, it is possible to produce electric vehicles in ways that minimise impact. But doing this will take time – which is why companies and governments need to make it a priority.

What companies can do

Car companies and their suppliers are the key to this, and there are three main ways they can help.

The first is to demand transparency from their suppliers – it needs to be easy to identify which company and mine provided which material, so carmakers can choose better suppliers, and steer clear of the bad ones.

Secondly car companies must only work with producers that follow the highest standards for workers' rights, pay and conditions and treatment of the environment.

Car companies are already huge buyers of materials like lithium cobalt and nickel, and that gives them serious influence over the mining industry that supplies them. They should make the most of this influence, offering long-term fixed contracts with these high standards written in.

Thirdly, these companies need to be willing to draw the line. Some things, like Indigenous rights, need to be sacrosanct and primary. And opening up already over-stressed oceans to seabed mining is a non-starter. That means some of these materials will simply have to stay in the ground.

What the government can do

Politicians also have an important role to play. They can:

- Exclude mining companies from the stock markets if they fail to meet the highest standards.
- Fund research on new battery tech that needs fewer mined materials.
- Regulate the industry to enforce 100% recycling of batteries.
- Set targets and take action to control and reduce air pollution from tyres.

The way forward

Electric vehicles are essential to meeting our climate targets, and that's why the government and industry need to get serious about making them better. But whether they're electric powered or fossil-fuelled, all cars put a burden on the environment. And when they're allowed to dominate urban areas, they make the streets hostile and dangerous to everyone else. So the UK also needs to look beyond swapping our petrol and diesel cars with an equivalent fleet of electric ones.

A transport policy fit for the current age wouldn't just focus on electric cars. It'd combine them with major investments in public transport, broadband, car sharing, walking and cycling. Cars have wreaked havoc on the climate and warped our physical world. It's time to push back on both fronts.

11 September 2020

The above information is reprinted with kind permission from GREENPEACE
© Greenpeace 2022

www.greenpeace.org.uk

Are smart solar flowers worth the recent hype?

Smart solar flowers or Smartflowers are large, mechanical blooms with solar panels for 'petals'. Some customers prefer this crafty invention over traditional rooftop panels. Here's why.

Smart solar flowers and trees are revolutionising the solar power industry by how they capture energy from the sun. The Smartflower features an array of panels, unfurling in the morning and rotating with the sun throughout the day. They fold up at night when their operation is unnecessary.

This mechanical device that looks like a large flower can capture the sun's rays and convert them into electricity. This innovative technology sounds like a worthy investment, but is it?

Solar flowers and trees are made for the purpose of decoration while they provide renewable energy for homes and businesses. However, many people considering the installation of solar panels wonder how practical these structures are for energy production. Here's everything you need to know about these powerful plants.

What can smart solar flowers and trees provide?

Smartflowers are solar installations mounted into the ground. It is a freestanding structure that's fully assembled upon delivery. The solar flowers have panels for petals and the trees feature branches with panels on top. Some solar trees are also built as a single tower with a photovoltaic panel mounted on top.

Solar flowers have 12 petals rotating at a 90-degree angle, producing 40% more energy than fixed solar panels. They work with a dual-axis tracking system, so they always face the sun — increasing productivity by up to 10%. Their elevated design also allows the panels to cool naturally, increasing their efficiency.

In addition, solar plants can benefit cities aiming to procure a green infrastructure. They can reduce energy usage and fossil fuel consumption when combining solar panels and natural elements. Essentially, a biosolar installation would transform an urban environment, resulting in a healthier and cleaner place to live. They also provide visual appeal and are better able to blend into their environments, appearing as just part of the landscape.

Smartflower vs traditional solar

Solar flowers and traditional systems both produce clean, renewable energy. However, they have several differences that are worth noting.

A Smartflower can produce 2.5 kW, equating to USD$10 per watt. On the other hand, traditional rooftop solar systems generate 4 kW, which comes down to USD$6.50 per watt.

Some traditional solar systems can also vary by around USD$3.50 per watt.

In comparison, the solar flower is much more expensive than a conventional solar roof panel. While the costs could be a setback for some, the solar flower does have quite a few perks that may offset the expense.

The pros and cons of solar plants

Solar flowers and trees have a self-cleaning system where debris falls off as the petals collapse at night. This feature is optimal since dirt can harden on stationary panels — making them more difficult to clean over time.

Moreover, a Smartflower installation takes only two to three hours, whereas rooftop solar installations can take up an entire day. Solar flowers are also easily movable and can be transported to different locations.

One drawback of the solar flower is that it needs a fairly large area for operation – at least 17 feet of clear space in diameter. Therefore, Smartflowers are the least ideal for large housing developments with small lots. However, solar trees are perfect for places that don't have much space – making this a more attractive option for neighbourhoods and urban environments.

The costs are typically greater since this technology is new, especially when there is a low supply. However, the price could eventually come down once demand and production increase. Those who anticipate these systems should monitor for a better price point.

Is the smartflower a worthy option?

A solar tree or flower sculpture can significantly impact your commitment to sustainability. Suppose you're willing to pay a premium for ground-mounted solar systems and have the space for it. That makes these installations a great option, especially if you aren't too concerned about meeting your entire home's electrical needs.

In addition, customers can receive a 26% federal solar tax credit. The technology can also make a great conversation piece, adding tasteful aesthetics to a homeowner's property. However, if you have practicality in mind, the Smartflower may not be a sensible choice.

While solar flowers can produce more energy at certain times than fixed arrays, this could be an advantage – especially as utilities add consumption and pricing in the early evening. There's also the added peace of mind to a ground mount. Solar roof panels can cost extra for reinstallation when roof repairs are needed.

Going solar with smartflowers

The Smartflower is a great choice for those who like the rotating panels and unique look. Furthermore, solar flowers and trees can create a visual statement for businesses showcasing their commitment to greener operations. While other economical options are available, solar flowers and trees could be worth the investment if you're willing to spend the upfront cost.

EO's Position: The world is rapidly losing sight of being able to stay under the 1.5C limit of global temperature rise to avoid the catastrophic impacts of climate change. We need to rapidly scale up renewable energy generation as quickly as possible. While smart solar flowers is still an emerging tech, with strong investment and inter-connected regional grids, solar has the potential to become the baseload generator of renewable energy in the US.

6 May 2022

The above information is reprinted with kind permission from EARTH.ORG
© 2022 Earth.Org

www.earth.org

High fossil fuel prices are good for the planet – here's how to keep them high while avoiding riots or hurting the poor

An article from The Conversation.

By Neil McCulloch, Associate Fellow of Political Economy, Institute of Development Studies

In the UK, it now costs more than £100 to fill up a typical family car with petrol, and oil prices could rise even further. But are such high prices for fossil fuels a bad thing? While attention is focused on measures to tackle the global cost of living crisis, there has been much less focus on a very uncomfortable truth – that solving the climate crisis requires fossil fuel prices for consumers to stay high forever.

Saying such a thing may seem tone deaf. Millions of households in rich countries are facing a choice between heating and eating. In poorer countries, the situation is immeasurably worse. Rising prices for gas have dramatically increased the cost of fertiliser, while the war in Ukraine is hampering the export of its wheat.

Together these are leading to spiralling food prices globally, triggering a surge in inflation and worsening the already dire food security situation in places such as Yemen, the Horn of Africa and Madagascar. We are already witnessing widespread foot riots just like those between 2008 and 2011, when citizens around the world protested the failure of their states to deliver their most basic right – the right to eat.

To mitigate the impact of high prices, we have seen a screeching reversal of energy policies around the world. In November 2021, governments at the COP26 climate conference in Glasgow pledged to tax carbon and eliminate fossil fuel subsidies. But faced with dramatic increases in the cost of fuel and electricity, those same governments have scrambled to slash taxes on energy, put in place price caps and introduce new subsidies.

Yet keeping global warming to under 1.5°C will require a dramatic reduction in the use of fossil fuels, starting now. The unfortunate reality is that one of the most effective ways of getting people to use less fossil fuel is to ensure they are expensive.

Of course, the best way of moving away from fossil fuels is for there to be better (and preferably cheaper) alternatives. But investment in these renewable alternatives will only happen if people are clearly switching to them, and that requires consumer prices for fossil fuels to remain high.

Fuelling riots

Of course, high fossil fuel prices are typically unpopular and can even lead to riots. Between 2005 and 2018, 41 countries had at least one riot directly associated with popular demand for fuel. In 2019 alone, there were major protests related to energy in Sudan, France, Zimbabwe, Haiti, Lebanon, Ecuador, Iraq, Chile and Iran – many of which turned into riots.

Colleagues and I recently published research showing that these riots are caused by price spikes, often after fuel subsidies have been removed. These price spikes triggered fuel riots when citizens felt they had no other options for voicing their anger over government policies and actions (or when states attempted to violently suppress them from doing so).

High prices, happy citizens

Is it possible to keep fossil fuel prices high without triggering riots? The key is to keep consumer prices high by increasing fuel taxes when international oil and gas prices do eventually fall. Making this politically acceptable requires two things to happen.

First, consumers will not accept high prices if it means high profits for fossil fuel companies. Maintaining high prices for consumers must be complemented by a radical overhaul of the taxation regime facing fossil fuel companies, not just one-off windfall taxes. Those taxes would maintain high consumer prices even though the fossil fuel companies wouldn't actually receive very much – enough to cover reasonable costs, but not enough to invest in further fossil fuel production. As the International Energy Agency has pointed out, to

achieve net zero by 2050, the amount of investment needed in new oil and gas production is zero.

Second, consumers will be much more willing to accept higher prices for fossil fuels if the additional tax they pay is returned to citizens as an equal carbon grant. Alaska has done something similar, putting a share of oil revenues into a 'permanent fund' which it then distributes through a cheque to every household each year (though this approach can go wrong – in Alaska politicians ended up cutting public services to maintain payments from the state fund).

Getting an annual payment, equal to the taxes imposed to keep fossil fuel prices high, would cushion the hurt from higher prices. It would also be progressive, since those who consume the most fossil fuels would pay more in tax, while those who consume little would pay less but receive the same payment from the fund and therefore end up in profit. There might also need to be additional compensation for poor groups with high fossil fuel usage, such as people on lower incomes who have to use their cars for work.

Soaring energy costs are a disaster for poor consumers worldwide. But ironically, they also provide an opportunity to shift the world from its fossil fuel addiction. If we take this chance to make fossil fuel prices permanently high, we can accelerate the transition to cleaner energy in a way that is fair for all, and avert deeper crises in the years ahead.

15 June 2022

The above information is reprinted with kind permission from The Conversation.
© 2010-2022, The Conversation Trust (UK) Limited

www.theconversation.com

A town in Devon now gets its gas supply from animal poo

A town in Devon has become the first in Britain to get all its gas supply from renewable energy – made from animal manure.

By Metro Science Reporter

Locals in South Molton in Devon get all their gas and nearly half their electricity from 'anaerobic digestion'.

The process sees energy crops, agricultural waste and animal manure transformed to generate renewable gas and electricity.

Its homes are being powered by renewable gas and electricity thanks to an Anaerobic Digestion (AD) facility located on the outskirts of the town.

South Molton's facility Condate Biogas uses crops and chicken poo to create enough renewable energy to provide gas to the entire town – population around 5,000.

The facility currently provides enough renewable gas to meet the needs of every single home – and also supplies 40% of the town's electricity.

The biogas gas is turned into renewable energy which is injected directly into the gas and electricity grid by the plant.

Everything Condate Biogas needs to generate renewable energy is supplied by local farmers.

Ixora Energy which run the plant say they are leading the way in helping the UK produce renewable energy in the midst of an energy crisis.

It says it is also helping the Government meet its Net Zero targets.

Darren Stockley, managing director for Condate Biogas, said: 'Thanks to operating efficiencies, we are now capable of generating enough renewable electricity to supply 70% of South Molton with renewable electricity and we could increase our gas output to cover over 3000 homes.

'This means we could supply a wider range of villages and neighbourhoods around South Molton.

'We hope to be in a position to be able to deliver the additional energy soon because we believe that locally produced renewable energy provides the key to solving the current UK energy crises.'

MP for North Devon, Selaine Saxby, said: 'It is incredible to think that South Molton's homes are being powered by renewable gas and electricity thanks to Condate Biogas's anaerobic digestion facility.

'There are very few towns in the UK that can claim to be powered by renewable energy and it is something we can all be proud of.

'It also makes fertiliser as a side product, which goes back on the local land growing the maize which in turn provides the matter to help produce the gas.

'A truly circular form of local, sustainable, energy production.'

The company, which employs six local people, has plans to increase its energy output in the near future in order to help deliver even more renewable energy for nearby communities.

16 March 2022

The above information is reprinted with kind permission from Metro & DMG Media Licensing.
© 2022 Associated Newspapers Limited

www.metro.co.uk

Energy Crisis

Chapter 2

Why is there an energy crisis?

While gas and oil companies report billions worth of profits, vulnerable households are faced with rising costs and impossible choices. Find out why we're facing an energy crisis, who's to blame and what we can do about it.

Why are energy bills rising?

In February 2022, energy regulator Ofgem announced it's raising the cap on the prices that energy companies are allowed to charge customers by 54%. That means millions of people across the UK will face steeper bills come April.

A price cap is a form of economic regulation that sets a limit on the prices that a utility provider can charge. For example, the price cap allows oil and gas companies to 'pass on all reasonable costs to customers, including increases in the cost of buying gas.'

There's a lot of misinformation over what's caused the hike in costs, but Ofgem itself states the increase in heating costs is because of a record rise in global gas prices.

Who's to blame for the UK's energy crisis?

Over the past decade, the UK government has failed to invest in insulation, despite calls from environmental and fuel poverty groups.

A whopping 87% of UK households rely on gas. Such overreliance means we're particularly vulnerable to changes in global prices, and that vulnerability is deepened by the fact our houses are poorly insulated compared to those in other European countries.

For decades successive governments, including this one, have missed opportunities to invest sufficiently in renewable energy. Had they prioritised clean energy and home insulation, the demand for gas to heat homes would be much lower and we'd be less exposed to current prices hikes.

The impact will be devastating, as more people face the choice of either buying food or heating their home. What's more, with price hikes expected to remain until next winter, health inequalities will widen as households are forced to keep heating switched off. The elderly and clinically vulnerable are most at risk, as colder homes can lead to:

- a higher chance of getting a respiratory infection and bronchitis
- stress on the cardiovascular system
- making asthma symptoms worse, or causing asthma to develop
- an increased risk of mental health problems.

No one should have to prioritise eating or heating, particularly when oil and gas giants are reporting their biggest profits in years. Urgent support is needed to help those most at risk.

What about switching to green energy?

At Friends of the Earth, we're always happy to advise our supporters on how to switch to green energy. We normally recommend that you check out our fantastic partners, Ecotricity or Good Energy. However, due to the ongoing energy crisis, now's not an ideal time to switch energy supplier.

How can we end the energy crisis?

The energy crisis should be a wakeup call for the government to finally do what it should've done years ago and cut ties with the unreliable fossil fuels. It's outrageous they've let it come to this, but it's now clear that we need a more reliable system to heat our homes, once and for all.

The government needs to:

- Urgently impose a windfall tax on oil and gas companies to help finance support for the most vulnerable households.
- Implement a longer-term strategy to move to more reliable energy sources and help people insulate their homes.

Together with dozens of other organisations, we've written an open letter urging Boris Johnson to act. Our next campaign, focused on home heating, will prioritise the needs of the population over the profits of oil and gas companies.

Ultimately, our vision is of a country where everyone has a warm home that's heated affordably and sustainable. We know it's possible.

10 February 2022

The above information is reprinted with kind permission from Friends of the Earth.
© 2022 Friends of the Earth Limited

www.friendsoftheearth.uk

British Gas owner's profits increase five-fold while energy bills soar

British Gas owner Centrica has seen its profits increase by five-fold while households worry about paying their bills this winter.

By James Hocakday

The company's operating profits in the six months to the end of June came to £1.34 billion amid a rampant rise in the cost of living.

It was significantly more than the £262 million recorded during the same period last year.

Meanwhile Shell has reported record profits of £9.4 billion, having doubled them in a single year.

The rise in profits comes as wholesale gas prices hit record highs across Europe this year, following Vladimir Putin's invasion of Ukraine.

Russia has halved supplies of natural gas via the Nord Stream 1 pipeline to the continent, causing further chaos in the market.

Although the UK is less dependent on Russian energy than mainland Europe, the move is still having a knock-on effect on the UK.

The cap on how much British households can be charged for energy is set to rise to £3,420 in October – potentially rising again to £3,850 in January.

People are already shelling out £1,971 per year on average for energy after the cap was raised in April – a rise of 54% – contributing to rising inflation.

Centrica paused dividends in 2020 and began major cost cutting, selling upstream assets and seeking to reinvent itself as an energy service provider.

'We've made significant progress de-risking the Group and building a stronger business for the benefit of all stakeholders,' CEO Chris O'Shea said in a statement.

The company said it would reinstate a progressive dividend, initially offering offer an interim payout of 1 pence per share.

Operating profits at British Gas fell 43% to £98 million during the first half of 2022 as it needed to buy more energy in the wholesale market than expected to cover customer demand.

It said it made £6 per customer profit after tax during the period. In May Centrica completed the sale of its 69% stake in Spirit Energy's Norwegian oil and gas assets to Norwegian private equity firm Sval Energi for around £560 million.

A government spokesperson said: 'Unlike Europe, Britain isn't dependent on Russian gas.

'The UK's secure and diverse energy supplies will ensure households, businesses and industry can be confident they can get the electricity and gas they need.

'However, we are vulnerable to volatile gas markets. While no national government can control the gas price, we have introduced an extraordinary £37billion package to help households, including £1,200 each for 8 million of the most vulnerable households.'

28 July 2022

The above information is reprinted with kind permission from *Metro* & DMG Media Licensing.
© 2022 Associated Newspapers Limited

www.metro.co.uk

How energy retail suppliers are supporting customers through the crisis

By Dhara Vyas

Energy retail companies supply heat, light and power to 28 million homes as well as every business in the country. The industry responded quickly to the challenges presented by the pandemic, adapting their operations and providing millions of pounds of extra support to customers while ensuring that supplies went uninterrupted.

However, in recent months, record spikes in the global wholesale price of gas have taken a heavy toll on the sector with 27 energy retail companies exiting the market since August 2021 – almost 2.3 million households have seen their supplier fail. The costs involved in the supplier of last resort process is reported to be around £2.6 billion to date – and is likely to increase. These costs will ultimately be picked up by customers.

Industry and consumer groups have been calling on Ofgem and the Government to improve regulation for many years – for example ensuring that companies are financially viable before they enter the market or being quicker to act when rules have been broken (for example when there are customer service issues). However, unprecedented, high global prices mean we've also seen well run companies going out of business.

Every country has been affected by the high prices, but the UK is particularly reliant on imported gas, using it to heat over 80% of our homes (which is exacerbated by the UK's draughty housing stock) as well as generating around 40% of our electricity through gas powered plants.

The price paid by domestic customers on standard variable – or default – tariffs (SVT) is capped by Ofgem. The current cap was set in August before the recent increases in wholesale costs – it doesn't reflect the actual costs suppliers are facing – and the unavoidable losses have proved too much for many of them to withstand. The price cap applies to domestic consumers, and we know business groups are worried about the prices paid by non-domestic energy customers too.

This issue has been further exacerbated as customers who were on cheaper fixed term deals have moved onto the price capped SVTs when their fixed deal expired (because they are by far the cheapest on the market). Even before the current crisis, the retail sector was in a fragile state with the largest energy suppliers operating at a loss. Most companies in this market are not able to withstand or absorb such sudden and extreme cost increases. The recent upheaval has prompted a very welcome (and long-overdue) review of the retail market and strengthening regulation to build the resilience of the market. However, with wholesale gas prices rising by over 500% in over a year, even well-run and financially responsible suppliers have struggled to cope.

As well as the mismatch between what they are paying and what they can charge, retailers have little room for

What can be done?

For years Energy UK has been calling for reform of the retail market, highlighting its fragility even before the spike in wholesale gas prices.

Ultimately any decisions that could help to ease or smooth costs in the coming months lie with the Government.

We regularly meet with parliamentarians through regular briefings and roundtables with politicians, and during this crisis have worked with industry and governments to advise on potenetial solutions.

Energy UK is calling for the UK Government to do more to support customers as energy bills continue to increase.

- Some options being discussed are:
- Remove VAT from Energy bills
- Move policy costs to general taxation
- Smooth the cost of failed suppliers (currently £2.6bn) over a number of years
- Some mechanism for smoothing wholesale costs – ensuring long term investment confidence is not compromised.

What are suppliers doing to help customers?

Energy suppliers have paid out hundreds of millions of pounds in emergency credit, payment plans and payment holidays since the beginning of 2020 to support household customers.

Winter commitments

16 suppliers, covering over 90% of customers have signed up to new winter commitments, which have and will see them improve the quality and availabilityof support they ofer to customers in need this winter – whether that be from financial difficulties, mental or physical health issues or other events.

The commitments were agreed between industry and the sectoral regulator, Ofgem and include pledges to increase awareness of the help available and make it easier for customers in financial difficulties to contact their energy company.

Warm Homes Discount

Under the Warm Home Discount (WHD) scheme, larger energy suppliers support people who are in, or at risk of, fuel poverty, by offering an annual £140 rebate on their electricity bill – in total £350m a year is provided. This is due to rise to £475m per year from 2022.

manoeuvre elsewhere as roughly 80% of the costs in an energy bill are out of their direct control – by far the biggest component is wholesale costs (around 40%).

The biggest worry for the industry is that while the current price cap is shielding customers from the record wholesale prices, when these feed into the next price cap, which comes into force in April, customers are likely to face a 50% increase in their bills (an extra £600–£700 a year for households with typical usage). We're in a position where energy suppliers are losing millions of pounds serving their domestic customers, and this is unsustainable in the long term.

Energy retail suppliers have provided millions of pounds of support for people who are struggling to pay their bills over the past two years – and will continue to do all that they can to help their customers. But this is an economy wide issue.

Inflation rose by 5.4% in December 2021 and there is little doubt that people are experiencing a cost-of-living crisis that is set to get worse with rising prices for food as well as other goods and services. We know that many people are going to struggle to make ends meet – and this includes paying their energy bills. While there is understandably a lot of concern for people who are already financially vulnerable, Energy UK is worried that millions more will be pulled into difficulty. These pressures are likely to continue to grow, with further price rises outstripping pay growth.

If Government doesn't act by April the impact of the cost-of-living crisis will be felt by the vast majority of households.

We're pleased Government is considering a whole suite of options to support help people so they can afford to pay their energy bills when the price cap is increased. Energy UK wants Ministers to ensure that the action they take:

- Delivers fair prices for customers
- Supports people in vulnerable circumstances (recognising this group may grow)
- Addresses any economy wide impacts of high energy prices
- Enables the investment in a secure, Net Zero energy system
- Enables a retail market which is investable, stable and innovating for Net Zero

In the long term it's in all our interests to ensure the energy market is a competitive successful market that attracts investment so that energy retail companies can innovate and deliver Net Zero.

2 February 2022

The information here reflects the state of play when the piece was written in February. Some information may have changed by the time of this book going to press. All up-to-date info is available on energy-uk.org.uk

The above information is reprinted with kind permission from Energy UK.
© 2022 Energy UK

www.energy-uk.org.uk

In an energy crisis, every watt counts. So yes, turning off your dishwasher can make a difference

An article from The Conversation.

By Anna Malos, Climateworks Centre - Country Lead, Australia, Monash University & Emi Minghui Gui, Climateworks Centre Energy System Lead, Monash University

Australia's east coast energy market has been on a rocky road for the past few weeks. It begs the question: how could the market change to avoid the next crisis?

To date, discussion has largely focused on the need to generate more energy. But there's another way to ease strain on the system – by using less energy.

Last week, New South Wales residents were asked to find safe ways to consume less power during the evening peak, such as not running dishwashers until after they went to bed. Such actions, when deployed at scale, can make a big difference to shoring up short-term supplies.

But Australia has only scratched the surface of what's possible when it comes to managing energy demand. As the transition away from fossil fuels continues, we should scrutinise every bit of electricity consumption to make sure it's needed. It's not about going without, but making the best use of what's available.

Getting smart about energy use

Asking people to reduce electricity use is known in energy circles as 'demand management'.

Sometimes it involves paying consumers to use less electricity. That's because offering financial rewards is far cheaper than blackouts or bringing more emergency reserve supply onto the market.

The current system of demand management is currently geared towards major energy consumers, such as industrial plants. AEMO has several mechanisms through which it pays big energy users to power down when the system is struggling.

But more can be done to encourage households to reduce their electricity demand.

Some energy retailers offer incentives to encourage households to reduce their use at given times. It might mean people turning down the heater, using appliances outside peak times or tapping into rooftop solar power stored in home batteries instead of taking power from the grid.

Householders signed up to the scheme are sent a text message asking them to propose a reduction in energy use ahead of an expected supply shortage. Credits are paid if the household achieves the reduction.

Reducing household electricity demand will become easier as home appliances become increasingly internet-enabled and remotely controlled. This allows people to, for example, turn off a home appliance while they're at work.

In future, it could even allow people to opt into a scheme where a retailer temporarily turns off appliances in thousands of homes when they're unoccupied.

Currently, only a small number of households take part in such schemes – but retailers see much greater potential. For instance, over the next four years Origin Energy proposes to scale up their scheme to 2,000 megawatts – capacity similar to a large power station such as Loy Yang A in Victoria.

Net-zero and beyond

There are many ways to improve the way we currently manage demand – and many of them can lead to lower bills for consumers.

Time-of-use tariffs, which offer cheaper electricity outside peak times, are a key potential measure. Some homes already use the lower overnight electricity rates to heat their hot water. But big energy users have traditionally made most use of these incentives.

As householders increasingly use smart meters – devices that digitally measure energy use – opting into these tariffs will become easier.

Appliances, lighting and heating connected to the internet can dramatically increase the broader power of demand management. Businesses could offer services to, for instance, monitor the wholesale electricity market and remotely turn on electric hot water heaters when prices are cheapest.

Managing energy demand is crucial for the longer-term transition to net-zero emissions. As sectors such as transport and industry become electrified or move to green hydrogen (produced by renewable energy), new supply challenges will emerge.

For heavy industry, reduced energy use – as part of a broader shift away from fossil fuels – will reduce business costs and increase competitiveness. A new report, which we contributed to, shows a coordinated transition could also lead to wider benefits such as thousands of new jobs and cuts to greenhouse gas emissions.

The challenge for AEMO is to integrate renewable energy generation and storage, and a far greater use of demand management, into its next plan for the national electricity market.

And much can be done at a household level. Millions of Australian homes are costly to heat or cool because they're poorly insulated and designed. All levels of government could support the proposed revision of the National Construction Code to increase energy performance standards.

Looking ahead

Managing demand makes sense well beyond a crisis. Doing it well will go a long way to creating the clean, affordable and reliable energy system Australians need.

The potential for demand management only grows as renewable energy makes the electricity system more decentralised, and technology enables consumers to participate more actively.

The Energy Security Board is taking the right steps by working on issues such as flexible demand and consumer technology choices. The next test is how well the nation's energy ministers embrace the power of managing energy demand.

20 June 2022

The above information is reprinted with kind permission from The Conversation.
© 2010-2022, The Conversation Trust (UK) Limited

www.theconversation.com

Martin Lewis 'ringing the alarm bell' on energy crisis as he receives CBE

The MoneySavingExpert founder said he felt he had to 'raise the alarm' because the country would face 'cataclysmic' problems in the winter.

By Genevieve Holl-Allen

Consumer expert Martin Lewis has warned that the energy crisis is 'potentially more dangerous to lives than the pandemic', and said he will write a briefing note to the Prince of Wales on the issue, as he received his CBE at Windsor Castle.

The broadcaster and founder of MoneySavingExpert.com spoke to Charles on Tuesday after the prince presented his honour.

Mr Lewis said he felt he had to 'raise the alarm' because the country would face 'cataclysmic' problems in the winter.

Speaking at Windsor, he added: 'I would be far happier to have come and got my honour and have a nice fun day and not be talking about this because the world is wonderful, but the world isn't wonderful right now. And I think this is potentially more dangerous to lives than the pandemic.

'It is a cataclysmic problem that is going to face the country this winter.

'I am without embarrassment, deliberately, provocatively, raising an alarm right now. And I will do that with everyone. And when you get the ear of the Prince of Wales for a moment like that, it seemed the right time to take advantage.'

Mr Lewis condemned the MPs running in the Tory leadership contest for their 'deafening silence' on how they would tackle the energy crisis on becoming prime minister in the autumn.

He said: 'We have a debate about tax at the moment. Tax is certainly an issue of political philosophy which I understand why that is important to the Conservative Party, but people need to be under no mistake, energy is not a totem, energy is the core of the real problem people will face.

'So it seems to me absolutely the first part of the debate that every one of these candidates should answer is: what are you going to do to help people with the catastrophe that is coming that, unless it is dealt with properly, will see people starve and freeze?

'The provisions that have been put in place so far help, but the increase in the prediction since May, when those provisions were put in, is £450. The help for most households is £400. So just the increase since May has already eaten up the help that will be coming.

'So the first question that I want to hear them answer is: what are you going to do about energy bills?

'Tax does not help everyone. It doesn't help those on the state pension, it doesn't help those on universal credit. I'll leave others to look at the inflation issue.

'So absolutely, for me, this should be front and centre of the agenda. And what I'm hearing at the moment is a deafening silence.'

> 'When you see a ship about to go into an iceberg, you don't go "phew", you ring the bloody alarm bell, and I'm ringing the alarm bell right now' – Martin Lewis

Asked how he remains motivated to continue to lobby the Government over the cost of living, he replied: 'If you had the number of desperate people telling you their stories every day, and you knew you had a voice that could be magnified because of the work I've done.

'I'm lucky enough that when I say things they tend to echo to the public, and I think there's some 400 MPs follow me on Twitter, so I know, when I say something, it might just reach ears that can do something.

'I think I would be abrogating my responsibility if I didn't use that voice at the moment.

'So even though I'm tired, and I'm going to take a bit of time off this summer, I feel it is a duty-bound responsibility. When you see a ship about to go into an iceberg, you don't go "phew", you ring the bloody alarm bell, and I'm ringing the alarm bell right now.'

12 July 2022

The above information is reprinted with kind permission from *The Independent*.
© independent.co.uk 2022

www.independent.co.uk

Facing up to the global energy crisis

The war in Ukraine has laid bare how energy resources are intertwined with political power and international security, plus how reliant we still are on fossil fuels to power our daily lives. This new podcast episode explores how we got to this point and what the future holds.

By Julie Weldon

In the latest episode in the WORLD: we got this podcast series, experts from our Faculty of Social Science & Public Policy look at global energy interdependence and what recent events reveal about how energy affects the social, political and economic order in Russia and the rest of the world.

The episode also looks at what China is doing to move away from its past reliance on coal to power its economic growth. And our experts explore what are the obstacles for many countries to transition away from fossil fuels onto renewable energy sources.

The revenue from the export of oil and gas that is usually referred to as 'rent' is a key source of income for Russia's federal budget - making up more than 30 per cent of it last year.

Kalina Damianova, a PhD candidate at King's Russia Institute and Graduate Teaching Assistant in the Department of Political Economy and Department of European and International Studies, says it has much wider implications too.

The rent has become a binding element in the overall social, political, and economic order in Russia. It alleviates some of the economic burden away from domestic consumers, it allows the political regime to fund social, political and economic projects and even benefits to some extent elite enrichment through these indirect sources of resource rent redistribution. So, in many aspects, energy resources in Russia are linked with power. – Kalina Damianova

She says that as things are still dynamic and changing, we can only speculate on the full implications of the war in Ukraine for Russia's role as an energy superpower. However, it is clear many countries will be reluctant to trust Russia in the future and its credibility as a reliable supplier is damaged.

One country that is continuing to take Russian oil and gas is China.

Isabel Hilton, a visiting professor at the Lau China Institute, says Beijing is unlikely to want to become too dependent on Russian energy. So, although we might see more Russian oil and gas going to China, it is unlikely to replace the European market.

She outlines how China powered much of its initial economic growth using its vast coal reserves, but it is moving away from that now and has committed to is reaching carbon neutrality by 2060 and for its emissions to peak by 2030 or earlier.

She says progress is slow and if China is to switch to more renewable energy sources, it will mean huge and complex changes to the energy grid and the market, especially as current contracts work against such a change.

However, she says China has initiatives to improve energy efficiency, has invested heavily in low carbon technologies, including electric vehicles, and lowered the cost of renewable technology for everyone. It has also said it will no longer build coal-fired power stations in other countries as part of its 'Belt and Road' strategy plus said it will support the development of renewable energy in host countries.

Isabel says we need more networks of cooperation and for businesses to move faster if we want to meet our energy needs at the same time as tackle the climate crisis.

We have the technology, we have the capability, but it demands political will to execute this change in our energy systems, and frankly I don't see the kind of leadership we need at the political level right now. So, I'm optimistic that it can be done. I'm optimistic actually about global public opinion… I'm optimistic that we can do it. I am just not so optimistic about whether we will do it. – Isabel Hilton

Dr Thomas Fröhlich, of our Department of War Studies, carries out work focusing on the geopolitical implications of the global energy transition.

In the episode he explores various alternatives to fossil fuels including economic degrowth, electrification, the bioeconomy (such as using ethanol as a fuel source which Brazil has tried) and hydrogen. In all cases, he says it is very difficult to get away from using fossil fuels to create energy because of the existing structures and processes we have in place designed around using gas and oil.

He outlines how economic growth could be combined with transitioning to alternative and sustainable energy sources. He also highlights things individuals can do and why it is vital that the financial industry and politicians move away from investing in or subsidising the fossil fuel industry. Looking to the future he does feel optimistic.

We already have seen a large shift in public opinion in favour of renewable energies…and this trend will only accelerate. So, while the pace at which we're going is not fast enough at this point, the path that we're following is the right one, and I am confident that within the next few years we will see things accelerate to the level needed. – Dr Thomas Fröhlich

11 May 2022

The above information is reprinted with kind permission from Kings College London.
© 2022 Kings College London

www.kcl.ac.uk

Breaking the gas habit: the energy crisis spotlights the need for green alternatives

Quick actions to help households use energy more efficiently would ease the current crisis and improve the UK's long term energy security.

By Madeleine Gabriel

Crises often present us with both challenges and opportunities – the current energy crisis is no different.

With the April price cap increase now in effect, the typical British household will be paying nearly £700 per year more for their energy – that's a jump of 54%. This is very bad news for households already struggling with rises in the cost of living.

It's estimated that energy bill increases will push an estimated 2 million more people into fuel poverty.

This huge and unwelcome leap is a result of spiralling wholesale gas prices. Energy suppliers could buy gas for their customers for around 65p per unit of energy last August, but were paying £2.70 per unit by January 2022. While gas is the culprit, the way energy markets work means that the price of electricity has gone up too.

A combination of factors has driven this surge. Demand for gas has grown globally as economies emerge from the COVID-19 pandemic, while Russia's invasion of Ukraine led to price spikes because of fears that supplies could be disrupted.

With gas prices so easily affected by shifts in demand and geopolitical events, it's hard to predict how they will change in future, but the outlook isn't good – it's widely expected that the price cap will increase again in October 2022.

In this context, switching from fossil fuels to green energy has quickly become even more important, not just to tackle climate change, but to make sure energy is affordable and reliable.

In the UK, we're highly gas-dependent. Around 40% of electricity comes from gas-fired power stations and 85% of households rely on gas for heating. But the writing is on the wall: if we want to have a long-term, tangible impact on fuel poverty, we need to invest in green alternatives.

The UK government has today published a new energy security strategy, aiming to cut reliance on imported fossil fuels. This includes a welcome focus on quickly ramping up offshore wind and solar electricity generation.

Increasing supply of green energy is just one side of the equation, though. We can also help break reliance on imports, reduce carbon emissions and cut costs for households by reducing the amount of energy we use.

Importantly, this doesn't have to come at a cost to comfort. We can reduce demand for energy by insulating homes and heating them with more efficient technologies. In the short term, optimising existing heating systems would be a big win. Although modern condensing boilers can be up to 98% efficient, most are running well below optimum efficiency.

Simple steps, such as turning down the flow temperature on combi boilers, could save households up to 8% of their gas usage. There's potential to go further too, by upgrading boilers so that they adjust their output depending on the outside temperature, and by using smart controls to help ensure homes only use the energy they need.

The real prize though will come from getting more electric heat pumps into homes.

Because they make use of ambient heat from the air or ground, heat pumps are much more efficient than gas boilers, producing 3-4 units of heat for every unit of electricity put in.

So why aren't we all using them already? In short, affordability. It costs on average £10.5k to install a heat pump into a home that's been using gas or oil for heating, and with electricity costing around four times as much as gas, heat pumps can be more expensive to run than gas boilers despite their greater efficiency.

Bringing down the cost of heat pumps is possible though. In fact, energy price spikes mean that for many the lifetime cost of a heat pump is already competitive with fossil fuel systems. Relatively small changes, such as removing green levies from electricity bills, could make heat pumps even more cost-effective, while the VAT cut announced in the Treasury's Spring Statement, and action announced today to boost UK heat pump manufacturing, could bring costs down further.

The energy costs crisis requires action to support vulnerable households right now, as well as action to shift quickly to greener, more efficient heating systems.

With energy so high in the public consciousness, there's a real opportunity to capitalise on demand for heat pumps and other efficiency measures.

We've found that up to 25% of consumers said they would choose a heat pump over a gas boiler at current prices – the current crisis could see those figures climb even higher.

What's crucial is that policies are implemented effectively. The new Boiler Upgrade Scheme in England and Wales, which launched on 1 April, must avoid the failures of the Green Homes Grant that preceded it, so that consumers can easily access the grants and installers aren't disadvantaged by delayed payments.

The current energy crisis is just that – a crisis. But it could also act as a catalysing moment – one that pushes us on to make difficult decisions about what our future looks like.

Ultimately, we don't have to remain tethered to the unpredictable fluctuations of the gas market. With green energy and technologies like heat pumps, we can invest in a more sustainable future for both people and the planet.

7 April 2022

The above information is reprinted with kind permission from Nesta.
© 2022 Nesta

www.nesta.org.uk

Building nuclear power stations in Scotland will help solve energy crisis, says Sir Keir Starmer

The Labour leader said the Scottish Government's opposition to nuclear energy meant the nation could miss out on millions of pounds of investment.

By Chris Green, Scotland Editor

New nuclear power stations should be built in Scotland to help solve the UK's energy crisis and bring household bills down, Sir Keir Starmer has said.

The Labour leader said the Scottish Government's total opposition to nuclear energy meant that the nation could miss out on millions of pounds of investment and thousands of jobs.

Sir Keir was speaking before a planned visit to Glasgow on Tuesday, where he will campaign alongside Scottish Labour leader Anas Sarwar before the local elections on 5 May.

Last week the UK Government's energy strategy included plans for eight new nuclear power stations, all of which will be built in England and Wales.

The Scottish Government has said it has 'no intention' of allowing any new reactors to be built north of the border and would focus instead on renewables.

Scotland currently has only one nuclear power station, the Torness plant in East Lothian, after the Hunterston B site in North Ayrshire was closed in January.

Sir Keir and Mr Sarwar said the SNP should reverse its opposition to nuclear in light of the energy crisis, arguing that it should be part of the UK's overall mix to ensure 'stability of output' and secure the country's future energy supply.

Labour also pointed out that by turning its back on nuclear, the Scottish Government could be jeopardising a bid for a state-of-the-art nuclear fusion plant in North Ayrshire.

The proposed facility at Ardeer would specialise in nuclear fusion rather than fission, with the potential to generate virtually unlimited supplies of low-carbon, low-radiation energy.

The UK Atomic Energy Authority has shortlisted Ardeer as a possible site for the UK's first nuclear fusion facility alongside four others, with a decision due later this year.

'We need answers that focus on bringing bills down long term, as well as meeting our commitments to cut our reliance on fossil fuels and make our energy supply more secure in an unstable world,' Mr Sarwar said.

'To do that nuclear, and the highly paid and skilled jobs it brings, must be part of Scotland's energy mix.

'But Scotland now risks paying the price in lost jobs and opportunities for the SNP's unscientific and economically backward opposition to nuclear energy.'

Sir Keir added: 'Both the SNP and the Tories need to get their act together when comes to dealing with the energy crisis.

'We need real investment in green and renewables jobs, not more broken promises.

'But we also need to seize the opportunities for investment and energy security that come with nuclear energy.'

Scottish Energy Secretary Michael Matheson said his government's opposition to new nuclear power stations was on environmental grounds, due to safety concerns and because 'it is probably the most expensive form of electricity you can choose to produce'.

11 April 2022

The above information is reprinted with kind permission from iNews.
© 2022 Associated Newspapers Limited.

www.inews.co.uk

Chapter 3

Future of Energy

What the future of renewable energy looks like

Renewable energy capacity is set to expand 50% between 2019 and 2024, led by solar energy. This is according to The International Energy Agency (IEA)'s 'Renewable 2020' report, which found that solar, wind and hydropower projects are rolling out at their fastest rate in four years, making for the argument that the future lies in using renewable energy.

By Emily Folk

The future of renewable energy: growth projections

Renewable energy resources make up 26% of the world's electricity today, but according to the IEA its share is expected to reach 30% by 2024. The resurgence follows a global slowdown in 2019, due to falling technology costs and rising environmental concerns.

Renewable energy in the future is predicted that by 2024, solar capacity in the world will grow by 600 gigawatts (GW), almost double the installed total electricity capacity of Japan. Overall, renewable electricity is predicted to grow by 1 200 GW by 2024, the equivalent of the total electricity capacity of the US.

The IEA is an autonomous inter-governmental organisation that was initially created after the wake of the 1973 oil crisis. It now acts as an energy policy advisor to 29 member countries and the European Commission to shape energy policies for a secure and sustainable future.

Solar will become 35% cheaper by 2024

When the sun shines onto a solar panel, energy from the sunlight is absorbed by the PV cells in the panel. This energy creates electrical charges that move in response to an internal electrical field in the cell, causing electricity to flow.

Industry experts predict that the US will double its solar installations to four million by 2023. In 2018, the UK had over one million solar panel installations, up by 2% from the previous year, and Australia reached two million solar installations in the same year. A big reason for this increased uptake is the fall in prices to install the panels.

The cost of solar PV-based power declined by 13% in 2018, while Carbon Tracker predicts that 72% of coal-based power will become globally unprofitable by 2040. The IEA report found that solar energy will account for 60% of the predicted renewable growth, primarily due to its accessibility. Compared with the previous six-year period, expansion of solar energy has more than doubled. The cost of solar power is expected to decline by 15% to 35% by 2024, spurring further growth over the second half of the decade.

Future capacity of solar energy

Wind and hydropower often require users to live in specific locations, but solar offers more freedom; the sun rises and sets on a predictable schedule, and it's not as variable as running water or wind. Residential solar power is expected to expand from 58 GW in 2018 to 142 GW by 2024, and annual capacity additions are expected to more than triple to over 20 GW by 2024. China is expected to register the largest installed residential solar capacity in the world by 2024, with the strongest per capita growth in Australia, Belgium, the Netherlands and Austria.

Solar facilities will continue reducing their variability rates by storing electricity during the day and running at night. However, advanced solar plants will operate on higher DC to AC ratios, meaning they'll deliver more consistent service for longer durations.

Commercial and residential buildings will keep running at full capacity even in periods of low sunlight. Closing the gaps between sunlight collection and electricity generation will spur residents and corporations to join the solar movement. Therefore, it's imperative for governments to implement incentive and remuneration schemes, as well as effective regulation policies. For example, California has mandated that after 2020, solar panels must be installed on new homes and buildings of up to three storeys.

Commercial and industrial solar energy capacity is forecast to constitute 377 GW in 2024, up from 150 GW in 2018, with China predicted to be the largest growth market. This market remains the largest growth segment because solar power is usually more inexpensive and has a relatively stable load profile during the day, which generally enables larger savings on electricity bills.

Onshore wind energy capacity will increase 57% by 2024

To generate electricity using wind, wind turns the propeller-like blades of a turbine around a rotor, which spins a generator, which creates electricity.

The adoption of wind power is becoming more prominent due to increased capacity.

Onshore wind capacity is expected to expand by 57% to 850 GW by 2024. Annual onshore wind additions will be led by the US and China, owing to a development rush and a policy transition to competitive auctions respectively. Expansion will accelerate in the EU as competitive auctions continue to keep costs relatively low. These auctions will mean that growth in Latin America, the MENA region, Eurasia and sub-Saharan Africa will remain stable over the forecast period.

Offshore wind capacity is forecast to increase almost threefold to 65 GW by 2024, representing almost 10% of total world wind generation. While the EU accounts for half of global offshore wind capacity expansion over the forecast period, on a country basis, China leads deployment, with 12.5 GW in development. The first large US capacity additions are also expected during the forecast period.

Japan expands wind energy

Japan is experimenting with the idea of installing offshore turbines to replace many of their nuclear reactors, a result of the country's 2011 nuclear disaster in Fukushima. The company Marubeni recently signed a project agreement to build offshore farms in northern Japan, with each farm able to produce 140 MW of power.

Japanese lawmakers have created regulations to give developers more certainty in constructing sources of wind-based electricity; legislation outlining competitive bidding processes has been passed to ensure that building costs are reduced and developers consider potential capacity issues. The country's Port and Harbour Law has also been revised to spur wind turbine construction in port-associated areas and other locations favourable to wind turbines.

Grid integration, financing and social acceptance remain the key challenges to faster wind expansion globally.

Hydroelectric capacity will rise 9% by 2024

Hydropower plants capture the energy of falling water to generate electricity. A turbine converts the kinetic energy of falling water into mechanical energy. Then a generator converts the mechanical energy from the turbine into electrical energy.

According to the IEA, hydropower will remain the world's primary source of renewable power in 2024. Capacity is set to increase 9% (121 GW) over the forecast period, led by China, India and Brazil. 25% of global growth is expected to come from just three megaprojects: two in China (the 16 GW Wudongde and 10 GW Baihetan projects) and one in Ethiopia (the 6.2 GW Grand Renaissance project).

However, there has been a slowdown in the two largest markets, China and Brazil; growth is challenged by rising investment costs due to limited remaining economical sites and extra expenditures in addressing social and environmental impacts.

Nevertheless, annual additions are expected to expand in sub-Saharan Africa and in the ASEAN region as untapped potential is used to meet rising power demand.

Geothermal capacity will increase 28% by 2024

To generate geothermal energy, hot water is pumped from deep underground through a well under high pressure. When the water reaches the surface, the pressure is dropped, which causes the water to turn into steam. The steam spins a turbine, which is connected to a generator that produces electricity. The steam cools off in a cooling tower and condenses back to water. The cooled water is pumped back into the Earth to begin the process again.

The US market for geothermal heat pumps will exceed $2 billion by 2024 as demand for efficient heating solutions increases. Transformed building codes will encourage a move to renewable heating and electricity systems in commercial and residential real estates.

Geothermal capacity is anticipated to grow 28%, reaching 18 GW by 2024, with Asia responsible for one-third of global expansion, particularly Indonesia and the Philippines, followed by Kenya, whose geothermal capacity is set to overtake Iceland's during the forecast period.

The same research from Global Market Insights predicts the commercial market will experience the most considerable uptick; according to the Department of Energy, geothermal solutions will generate 8.5% of all electricity in the US by 2050.

The future lies in using renewable energy

Renewable energy will continue to rise in the upcoming decade, edging out fossil fuels and reducing greenhouse gas emissions.

'This is a pivotal time for renewable energy,' said the IEA's executive director, Fatih Birol. 'Technologies such as solar and wind are at the heart of transformations taking place across the global energy system. Their increasing deployment is crucial for efforts to tackle greenhouse gas emissions, reduce air pollution, and expand energy access.'

22 August 2021

The above information is reprinted with kind permission from EARTH.ORG
© 2022 Earth.Org

www.earth.org

World's first sand-powered battery is ready to heat homes

Researchers say the Finnish innovation could help solve the problem of how to store renewable energy such as wind and solar.

By Emma Gatten, Environment Editor

A battery that uses sand to store heat for months and could help with the green energy transition has been unveiled in Finland in a world first.

The sand battery will be used to provide heating to homes in the city of Kankaanpää in Western Finland, as well as its local swimming pool.

Researchers say it could help solve the problem of how to store renewable energy such as wind and solar as it can be used when the sun is not shining or the wind is not blowing.

The sand, which can reach temperatures of 600C, provides an alternative to lithium-ion batteries, which rely on expensive and relatively scarce metals and can only store a limited amount of energy.

The initial version installed in Kankaanpää uses 100 tons of sand in a seven-metre-high steel container to store energy on a site at the local power plant which will provide district heating to around 100 homes in the area throughout the year.

Eventually, the researchers plan to scale up to create batteries around 100 times that size, which would be able to take advantage of sunny or windy periods to store up energy.

'This innovation is a part of the smart and green energy transition,' said Markku Ylönen, the co-founder of Polar Night Energy, which came up with the sand battery.

Other countries are working on similar sand battery projects, including the American National Renewable Energy Laboratory, but Polar Night Energy says it is the first fully operational commercial scale project.

Sand is the world's second most used resource, and facing growing shortages, but Mr Ylonen said the battery storage would use 'low value' and 'sand like' substances, to allay concerns about running out of raw materials.

The final major challenge for the project will be to use the energy in the sand to discharge electricity, rather than simply heat.

Although it can be done, with current technology the process is inefficient and therefore too expensive to be done at scale.

One of the best uses of the sand battery may be to provide energy for industrial processes that require a lot of heat, said Mr Ylonen.

5 July 2022

The above information is reprinted with kind permission from *The Telegraph*.
© Telegraph Media Group Limited 2019

www.telegraph.co.uk

Shell to build Europe's biggest renewable hydrogen plant

♦ Shell will start building new renewable hydrogen plant in the Netherlands
♦ Holland Hydrogen I to produce 60k kilograms of renewable hydrogen a day

By Jane Denton

Shell is to start building a renewable hydrogen plant in the Netherlands, which is slated to be Europe's largest once operational in 2025.

The energy giant said the 200 megawatt electrolyser, called 'Holland Hydrogen I', in the port of Rotterdam, will produce up to 60,000 kilograms of renewable hydrogen a day.

The London-based group, which aims to become a net zero greenhouse gas emissions company by 2050, has been boosting its low-carbon output as it shifts away from oil and gas.

Shell said it aims to produce hydrogen at the plant using electricity generated by the offshore wind park, Hollandse Kust Noord, which it partly owns.

The renewable hydrogen produced will supply the Shell Energy and Chemicals Park Rotterdam, by way of the HyTransPort pipeline, where it will replace some of the grey hydrogen usage in the refinery.

Anna Mascolo, executive vice president of Shell's emerging energy solutions, said: 'Holland Hydrogen I demonstrates how new energy solutions can work together to meet society's need for cleaner energy.

'It is also another example of Shell's own efforts and commitment to become a net-zero emissions business by 2050.

'Renewable hydrogen will play a pivotal role in the energy system of the future and this project is an important step in helping hydrogen fulfil that potential.'

Renewable and low-carbon hydrogen is crucial in curbing emissions, but it will only account for 5 per cent of the global final energy mix by 2050, falling short of what is needed to meet climate goals, global energy consultancy DNV said last month.

Earlier this week Shell became the latest company to join the world's biggest liquefied natural gas project in Qatar.

The group will take a 6.25 per cent stake in the £24 billion North Field East scheme run by Qatar Energy.

Shell was the final energy company to partner with Qatar in the project following earlier deals with Total Energies, Exxon, Conoco Phillips and Eni.

The expansion of the scheme will boost Qatar's position as the world's top LNG exporter as demand for the fuel soars following Russia's invasion of Ukraine.

Shell shares were up 1.76 per cent or 35.50p to 2,051.50p this morning, having jumped by around 40 per cent in the last year.

Oil prices increased by around 3 per cent to over $100 a barrel again before paring some gains today, as investors piled back into the market after a heavy rout in the previous session, with supply concerns returning to the fore even as worries about a global recession linger.

Russ Mould, investment director at AJ Bell, said: 'The recent slump in the oil price is bad news for some of the biggest names on the UK market, principally BP and Shell.'

He added: 'Having slumped in price last night, oil managed to claw back some of its losses on Wednesday as the market stabilised. The question is, how long will this stability last? On one hand, a recession could easily reduce oil demand. On the other, supplies remain tight, so we perhaps won't see a big price crash if the world grinds to an economic halt.

'Wednesday's oil price bounce back helped lift the FTSE 100 by 2.4 per cent and gave support to markets across the rest of Europe. However, don't be fooled into thinking this is the start of a big recovery. Markets are likely to stay volatile for the near-term.'

6 July 2022

The above information is reprinted with kind permission from *This is MONEY* & DMG Media Licensing.
© 2022 This is Money

www.thisismoney.co.uk

The English city using community energy to drive positive change

From urban hydropower to microgrids and rooftop solar, Bristol Energy Cooperative harnesses green energy and the power of community to target a self-sufficent future.

By Duncan Jeffries

Bristol is no stranger to solar panels. Around 30MW of solar PV has already been installed on rooftops across the city, allowing businesses, schools, community centres and residents to benefit from renewable energy. But according to Bristol's Centre for Sustainable Energy, that's a mere 6 per cent of what's possible.

Working to increase that figure is Bristol Energy Cooperative (BEC). It provides organisations with free solar panels, and supports on a myriad of projects that are expanding the city's green energy supplies.

Since 2011, BEC has raised £14 million through bond and share offers, and loans. It's installed over 9MWp (mega watt peak refers to the maximum potential output of power) of solar and battery assets – enough to power more than 3,000 homes – and channelled over £300,000 of community benefit payments into local social and environmental initiatives. It's not bad going for an organisation that was established by volunteers. People who, as communications manager Jess Gitsham puts it, 'just wanted to take some action to address climate challenges'.

Today, BEC is in the midst of its eighth share offer. Initially aiming for a £1 million target, in April they took the decision to double this to £2 million, and are now about halfway to that goal. The money will fund community-owned green energy projects in Bristol and each investor, regardless of their investment, is granted an equal say in how the organisation is run. Projected returns are 3.5 per cent per annum, a rate that's a ways off from current levels of inflation, but, says, Gitsham: 'The 'added value' of our investment v an ESG fund [one incorporating environmental, social and governance criteria] is that it allows people to learn and engage in the work – to have a much closer connection with how their money is making a difference.'

The purse is powerful when it comes to generating change, she believes. As opposed to pension funds, which can leave you feeling like an insignificant cog in a less than benevolent wheel, community energy is about taking control of your savings. It's about 'channelling them into real-life change,' she says.

One project the share offer will help to fund is a 125kW rooftop PV scheme for Bristol Beacon (formerly Colston Hall), which has set its sights on becoming the UK's first carbon-neutral concert hall. BEC's involvement has transformed the scale of the plans, says Will Houghton, the co-op's project developer. 'It was going to be a small installation on one side of the roof, and we're covering the whole roof now,' he explains.

BEC will also provide funding for a community solar PV microgrid for Bridport Cohousing's 'sustainable and affordable' 53-home development in Dorset. The homes will have air source heat pumps for hot water and back-up heating, and PV panels on their roofs. Energy from these will be linked to the on-site microgrid. It is the largest project of its kind in the UK, and a model that could be adopted by other housing developers.

'We're hoping that after a few years of operation, the Bridport Cohousing microgrid can be passed over into ownership by the residents,' says Houghton.

A previous share offer also raised funding for the Tesla battery and microgrid that is now installed at the Water Lilies housing development in the Lawrence Weston area of Bristol – the UK's first domestic housing microgrid with battery storage.

An urban hydroelectric scheme at Netham Weir on the River Avon is in the works too, which would provide around 300kWp – enough to power around 250 homes.

'We're very excited about it because there aren't many opportunities for hydro in Bristol,' says Houghton. At the moment, most of the energy produced by water crashing over the weir is lost. 'Standing next to it, you can really sense how much energy there is,' Houghton enthuses. 'We're trying to do something with that – to make sure it's not wasted.'

By investing in community energy projects, anyone can make a difference to the way energy is produced in their local area – one they can see with their own eyes. This is particularly powerful at a time when many people are concerned about rising energy costs and progress on net zero targets.

One of BEC's investor-members recently told the organisation about a visit to Netham Weir with her three children. 'She loved the fact that she could go into our city and see this stuff going on, and that she's been involved in the investment behind it,' says Gitsham, who hopes such projects will inspire the next generation of engineers.

There are now 424 community energy organisations across the UK, according to Community Energy England's 2021 state of the sector report. 'That's pretty massive, but not that many of us hear about it,' says Gitsham. She advises anyone who wants to get involved in community energy to get in touch with their local cooperative.

'Time and resources are always a challenge. So willing volunteers who can bring their skills along, whatever they are, are always welcome.'

Bristol Energy Cooperative in numbers

The facts:

£14 millon raised since 2011

9MWp of solar and battery assets currently installed

3,000 number of homes powered by those assets

3.5 % for investors

1 June 2022

The above information is reprinted with kind permission from *Positive News* in association with Good Energy.
© 2022 Positive News Publishing Limited

www.positive.news

Renewable energy – powering a safer future

Energy is at the heart of the climate challenge – and key to the solution.

A large chunk of the greenhouse gases that blanket the Earth and trap the sun's heat are generated through energy production, by burning fossil fuels to generate electricity and heat.

Fossil fuels, such as coal, oil and gas, are by far the largest contributor to global climate change, accounting for over 75 percent of global greenhouse gas emissions and nearly 90 percent of all carbon dioxide emissions.

The science is clear: to avoid the worst impacts of climate change, emissions need to be reduced by almost half by 2030 and reach net-zero by 2050.

To achieve this, we need to end our reliance on fossil fuels and invest in alternative sources of energy that are clean, accessible, affordable, sustainable, and reliable.

Renewable energy sources – which are available in abundance all around us, provided by the sun, wind, water, waste, and heat from the Earth – are replenished by nature and emit little to no greenhouse gases or pollutants into the air.

Fossil fuels still account for more than 80 percent of global energy production, but cleaner sources of energy are gaining ground. About 29 percent of electricity currently comes from renewable sources.

Here are five reasons why accelerating the transition to clean energy is the pathway to a healthy, livable planet today and for generations to come.

1. Renewable energy sources are all around us

About 80 percent of the global population lives in countries that are net-importers of fossil fuels – that's about 6 billion people who are dependent on fossil fuels from other countries, which makes them vulnerable to geopolitical shocks and crises.

In contrast, renewable energy sources are available in all countries, and their potential is yet to be fully harnessed. The International Renewable Energy Agency (IRENA) estimates that 90 percent of the world's electricity can and should come from renewable energy by 2050.

Renewables offer a way out of import dependency, allowing countries to diversify their economies and protect them from the unpredictable price swings of fossil fuels, while driving inclusive economic growth, new jobs, and poverty alleviation.

2. Renewable energy is cheaper

Renewable energy actually is the cheapest power option in most parts of the world today. Prices for renewable energy technologies are dropping rapidly. The cost of electricity from solar power fell by 85 percent between 2010 and 2020. Costs of onshore and offshore wind energy fell by 56 percent and 48 percent respectively.

Falling prices make renewable energy more attractive all around – including to low- and middle-income countries, where most of the additional demand for new electricity will come from. With falling costs, there is a real opportunity for much of the new power supply over the coming years to be provided by low-carbon sources.

Cheap electricity from renewable sources could provide 65 percent of the world's total electricity supply by 2030. It could decarbonize 90 percent of the power sector by 2050, massively cutting carbon emissions and helping to mitigate climate change.

Although solar and wind power costs are expected to remain higher in 2022 and 2023 then pre-pandemic levels due to general elevated commodity and freight prices, their competitiveness actually improves due to much sharper increases in gas and coal prices, says the International Energy Agency (IEA).

3. Renewable energy is healthier

According to the World Health Organization (WHO), about 99 percent of people in the world breathe air that exceeds air quality limits and threatens their health, and more than 13 million deaths around the world each year are due to avoidable environmental causes, including air pollution.

The unhealthy levels of fine particulate matter and nitrogen dioxide originate mainly from the burning of fossil fuels. In 2018, air pollution from fossil fuels caused $2.9 trillion in health and economic costs, about $8 billion a day.

Switching to clean sources of energy, such as wind and solar, thus helps address not only climate change but also air pollution and health.

4. Renewable energy creates jobs

Every dollar of investment in renewables creates three times more jobs than in the fossil fuel industry. The IEA estimates that the transition towards net-zero emissions will lead to an overall increase in energy sector jobs: while about 5 million jobs in fossil fuel production could be lost by 2030, an estimated 14 million new jobs would be created in clean energy, resulting in a net gain of 9 million jobs.

In addition, energy-related industries would require a further 16 million workers, for instance to take on new roles in manufacturing of electric vehicles and hyper-efficient appliances or in innovative technologies such as hydrogen. This means that a total of more than 30 million jobs could be created in clean energy, efficiency, and low-emissions technologies by 2030.

Ensuring a just transition, placing the needs and rights of people at the heart of the energy transition will be paramount to make sure no one is left behind.

5. Renewable energy makes economic sense

About $5.9 trillion was spent on subsidizing the fossil fuel industry in 2020, including through explicit subsidies, tax breaks, and health and environmental damages that were not priced into the cost of fossil fuels.

In comparison, about $4 trillion a year needs to be invested in renewable energy until 2030 – including investments in technology and infrastructure – to allow us to reach net-zero emissions by 2050.

The upfront cost can be daunting for many countries with limited resources, and many will need financial and technical support to make the transition. But investments in renewable energy will pay off. The reduction of pollution and climate impacts alone could save the world up to $4.2 trillion per year by 2030.

Moreover, efficient, reliable renewable technologies can create a system less prone to market shocks and improve resilience and energy security by diversifying power supply options.

Article accessed 7 July 2022: https://www.un.org/en/climatechange/raising-ambition/renewable-energy
The above information is reprinted with kind permission from United Nations.
© 2022 United Nations

www.un.org

Powered by poo? The weird and wonderful alternative energy sources that could soon be powering your home

By Ben Gallizzi, Energy Expert

The UK government has set a world-leading net-zero target, the first major economy to do so. Tackling climate change will require decisive global action and the way we produce and use energy is at the heart of this debate. It's no surprise then that the need for alternative energy sources is so crucial.

Luckily, there are scientists across the globe, and right here in the UK, who are trialling some of the more unconventional, ridiculous, and, in some cases, unsavoury solutions to abundant and cheap energy sources. Could dancing in our kitchens power our kettles in five years? Or the waste from our bathroom power our shower? It's time to think outside the box.

These alternative energy sources are not too far off reality, according to recent research compiled by the energy experts at money.co.uk. Which could mean big changes to what consumers look for when searching for energy deals.

We've consulted with several energy experts globally to compare the ten sources of renewable energy that are likely to be powering our homes in the not-so-distant future.

1. Poo power - Kingston Upon Thames, UK

Otherwise known as biogas energy, 'poo power' is the result of excess heat recovered from the sewage treatment process - delightful, we know. This odd yet unique type of energy source is predicted to work wonders by powering 2,000 new homes on Kingston's Cambridge Road Estate - all thanks to a new carbon-cutting partnership between Thames Water and Kingston Council.

The 'poo-power' energy scheme has the potential to provide clean, green heating to new homes and, if successful, is expected to be a model for similar schemes across the UK. With the ability to reduce millions of tonnes of carbon emissions each year, this innovative green energy plan is definitely one to watch.

Sarah Bentley, Thames Water CEO, said: 'Not only do we provide life's essential service, clean and fresh drinking water to millions of customers every day, but we also create reliable, affordable, and sustainable power by processing sewage. For us, the next stop is net zero.

'Achieving this target will require us to explore innovative new solutions and technology, led by our net-zero task force of experts from across the business. We don't yet have all the answers and our plans will evolve, but it's a challenge we're all relishing to enable our customers, communities, and the environment to thrive.'

2. Kinetic energy floors - Rotterdam, Netherlands

Are your moves on the dance floor truly electrifying? Ever dreamed of a disco dancing session that could power your kettle? Look no further, for kinetic floor tiles that capture the energy generated from footfall have now hit the market.

Dutch artist and innovator Daan Roosegaarde and his team of designers and engineers have created an interactive dance floor for a nightclub in Rotterdam that generates electricity through the act of dancing. The sustainable dance floor produces up to 25 watts per module, meaning that the generated energy can be used to power the lighting and DJ booth. Who knew going green could be groovy?

3. Pee power - Bristol, UK

If number twos can power our homes, then why not number ones? Pee power (or, more formally, Microbial Fuel Cell energy system) has been developed by Bristol University and is set to be trialled for the first time in a residential setting this year. The property in question will be the only home in the world to use 'pee power' as an energy source for electricity.

Bristol BioEnergy Centre Director Professor Ioannis A. Ieropoulos, who worked on the trial, comments: 'There is a real desire for schemes like ours, we've had generally very positive feedback as people seem to be open to the prospect of converting waste directly into electricity. Every time we demonstrate a successful installation, we get closer to wider-scale implementation, but for this to become the norm, it will need to form part of new planning and building legislation.'

4. Cocoa - Ivory Coast, West Africa

After successful pilot projects, the Ivory Coast has begun work this year on a biomass plant that will run on cocoa waste. The facility will be based in Divo, a town that produces a large share of the country's cocoa.

Cocoa plant matter left over after cocoa production will be burned in the biomass plant, helping to turn a turbine and generate electricity, much like a conventional fossil fuel power plant. Apparently cocoa does have the power to give us more than a sugar kick.

'This plant alone will be able to meet the electricity needs of 1.7 million people,' says Yapi Ogou, managing director of the Ivorian company Société des Energies Nouvelles (Soden), which is involved in building the plant.

5. Solar roads - Germany

Scientists in Germany and Austria think they have found a new energy use solution: a solar highway. This invention features solar panels on a canopy over existing highways to help supply green power to the nation.

There's a three-year plan in place to explore how they can cover the 13,000km highway network, but based on the power consumption of Germany in 2019, a solar highway would cover 9% of total power consumption. That's equivalent to 1/3 of the energy needed to power each and every home.

6. Crematorium, UK

Some would call this one dead heat. In the UK, 79% of people are now cremated, which is around 470,000 people each year spread across the UK's 300 crematoriums. Though gruesome at first glance, this type of energy source truly produces 'life after death'.

In the UK alone, significant power has been generated via crematoriums. A single cremation alone is to power 1,500 televisions.

A crematorium in Durham was the first back in 2011 when they installed turbines, which used heat generated during cremation. The turbines run on the steam that comes from cooling the hot gases used for cremation. The crematorium is then able to sell off electricity to the National Grid and use a third burner to heat their offices and chapel. Since then we've seen the Aalborg crematorium in Denmark make money by selling heat to nearby villages, whilst another in Redditch, warms the waters of a nearby swimming pool with the heat it produces.

7. Jellyfish, Sweden

With more power than just a swift sting, jellyfish could become the next big renewable energy source. The secret lies in a jellyfish's green fluorescent protein (GFP), giving some jellyfish their eerie glow. This substance reacts to UV light and excites electrons.

If silicon in solar panels can be replaced with jellyfish GFP, this energy-consuming process could be lessened in the future. Zackary Chiragwandi at Chalmers University of Technology in Gothenburg, Sweden, and colleagues are developing an electrical device based on green fluorescent protein (GFP) from the jellyfish Aequorea victoria.

8. Seaweed - Scotland, UK

Seaweed, you say? That's right - a car powered by Scottish seaweed has completed a 50-mile journey as part of an international project to develop greener fuels.

The vehicle set off from the Danish Technological Institute in the city of Aarhus and took a test drive on typical city roads and motorways to allow scientists to see how it performed.

It was part of an EU-funded project called MacroFuels, which has been developing cleaner alternatives to standard petrol and diesel by making biofuels derived from seaweed and algae.

9. Body heat - America

One of the biggest malls in America (it has its own zip code), the Mall of America in Minnesota is warmed in part by recycled heat from its shoppers.

70 degrees is maintained year-round with passive solar energy from 1.2 miles of skylights and heat generated from lighting, store fixtures, and body heat.

10. Cows wearing backpacks - Argentina

Methane from cows, which also comes from landfills and coal mines, is 28 times more powerful than carbon dioxide at trapping heat in the atmosphere. It's why researchers in Argentina attached balloon-like plastic packs to the backs of ten cows. Each pack had a tube from the animal's stomach that collected the gas.

Whilst the backpacks are a bit of a novelty, there is a very real opportunity to generate electricity from methane digesters on dairy farms.

Setting out to find the most interesting energy trials and schemes from the world's best scientists has truly been revealing. We were fascinated to find out that dance floor kitchen tiles could soon power your oven or that jellyfish solar panels could be lining our rooftops.

Imagine what our energy comparison sites will look like in the future and how we might compare the best energy deals.

28 July 2021

The above information is reprinted with kind permission from Money.
© Dot Zinc Limited 2022

www.money.co.uk

Key Facts

- 185 countries have signed the 2016 Paris Agreement. (page 1)

- France, Britain, and Germany (among others) have committed to getting rid of all their coal-fired plants by 2021, 2025, and 2038 respectively. (page 1)

- For more than two weeks in summer 2019, Britain didn't use any coal to generate electricity – the longest such period since 1882. (page 1)

- According to NASA, humans have increased atmospheric CO_2 concentration by more than a third since the Industrial Revolution began. (page 1)

- Air pollution resulting from coal kills 800,000 people per year. (page 3)

- The UN has stated that a sea level rise of just 0.5m could displace a total of 1.2 million people from low-lying islands in the Caribbean Sea, Indian Ocean, and Pacific Ocean. (page 3)

- Hydropower is the most widely used form of renewable energy in the world, producing 1,295 gigawatts of energy. This amounts to 54% of the global renewable power generation capacity. (page 5)

- Wind energy is the second most used renewable energy source in the world, producing 563 GW, and produces 24% of the world's total renewable energy generation capacity. (page 5)

- Iceland is one of the world's biggest producers of geothermal electricity, producing 26.5% of the country's electricity and 87% of their housing and building needs from natural hot water sourced underground. (page 5)

- The world's largest wind farm, Hornsea 2, is located 89km off the Yorkshire coast, UK. It spans an area of 462 sq.km – equal to 64,000 football pitches. (page 6)

- 2020 marked the first year in the UK's history that electricity came predominantly from renewable energy, with 43% of our power coming from a mix of wind, solar, bioenergy and hydroelectric sources. (page 8)

- The latest figures show there are now more than 10 million electric cars on the road globally, along with about one million electric vans, heavy trucks and buses. (page 14)

- British Gas's operating profits in the six months to the end of June 2022 came to £1.34 billion. (page 21)

- Energy retail companies supply heat, light and power to 28 million homes as well as every business in the country. (page 22)

- It's estimated that energy bill increases (in 2022) will push an estimated 2 million more people into fuel poverty. (page 28)

- In the UK, we're highly gas-dependent. Around 40% of electricity comes from gas-fired power stations and 85% of households rely on gas for heating. (page 28)

- Renewable energy resources make up 26% of the world's electricity today, but according to the IEA its share is expected to reach 30% by 2024. (page 30)

- There are now 424 community energy organisations across the UK, according to Community Energy England's 2021 state of the sector report. (page 36)

- Fossil fuels, such as coal, oil and gas, are by far the largest contributor to global climate change, accounting for over 75 percent of global greenhouse gas emissions and nearly 90 percent of all carbon dioxide emissions. (page 35)

- About 80 percent of the global population lives in countries that are net-importers of fossil fuels – that's about 6 billion people who are dependent on fossil fuels from other countries, which makes them vulnerable to geopolitical shocks and crises. (page 36)

- Methane from cows, which also comes from landfills and coal mines, is 28 times more powerful than carbon dioxide at trapping heat in the atmosphere. (page 39)

Glossary

Biofuel
A gaseous, liquid or solid fuel derived from a biological source, e.g. ethanol, rapeseed oil. Some scientists claim that GM would be a useful tool in the quest to produce biofuels which would be beneficial for the environment.

Biofuels and biomass
Plants use photosynthesis to store energy from the Sun in their leaves and stems. Living things, like these plant materials, are known as biomass. The wide range of fuels derived from biomass are known as biofuels. Corn ethanol, sugar ethanol and biodiesel are the primary biofuels markets.

Energy
A force which powers or drives something. It is usually generated by burning a fuel such as coal or oil, or by harnessing natural heat or movement (for example by using a wind turbine).

Fossil fuels
Fossil fuels are stores of energy formed from the remains of plants and animals that were alive millions of years ago. Coal, oil and gas are examples of fossil fuels. They are also known as non-renewable sources of energy, because they will eventually be used up: as they are finite, once they are gone we will be unable to produce more of them.

Geothermal power
The Earth is hot inside. Most of this heat comes from radioactive decay, which heats up the surrounding rocks. This heat can be used as an energy source: water is pumped down into the hot rocks, and the steam produced used to drive an electricity generator. The hot water can also be used directly to heat homes and businesses. However, unless the rock conditions are just right, geothermal power is not cost-effective.

Greenwashing
'Greenwashing' occurs when organisations falsely promote or market themselves as having 'green', environmentally-friendly, practices.

Heat pump
A device that captures heat from the air or ground and moves it into your home. It uses electricity to work but the amount of heat it harnesses is much greater than the amount of electricity used to operate it.

Hydropower
Energy which is generated using the movement of running water. This includes tidal/wave power.

Non-renewable energy
Energy generated from finite resources such as fossil fuels, energy can be generated from these sources, however, they will eventually run out.

Nuclear power
A method of generating energy using controlled nuclear reactions. These are used to create steam, which then powers a generator. Nuclear power is controversial and subject to much debate, with proponents saying it is a greener and more sustainable alternative to fossil fuels, whereas opponents argue that nuclear waste is potentially hazardous to people and the environment.

Offshore wind farm
An offshore wind farm consists of a number of wind turbines, constructed in an area of water where wind speeds are high in order to maximise the amount of energy which can be generated from wind.

Renewable energy
Energy generated from natural resources such as wind, water or the Sun. Unlike fossil fuels, energy can be generated from these sources indefinitely as they will never run out.

Solar power
Energy generated by harnessing the heat of the Sun.

Tidal power
A renewable source of energy generated by the surge of ocean waters during the rise and fall of tides.

Wind power
Energy which is generated using movement powered by wind. This is most commonly achieved via wind turbines, which are used to produce electricity.

Activities

Brainstorming

- As a class, discuss what you know about energy. Consider the following points:
    - What are fossil fuels?
    - What is renewable energy?
    - What are carbon emissions?
    - What does 'net zero' mean?
    - What does it mean to be 'energy efficient'?
    - Where does the energy that powers the UK come from?
    - What are the 4 main renewable energy sources used to power the UK?
- In small groups, discuss what kind of measures people can take in their own homes to be more energy efficient.

Research

- Conduct a survey among the class to find out how many different types of energy sources and energy efficiency measures are used in their households. Consider asking if they have:
    - Solar panels
    - Double glazing
    - Loft insulation
    - Electric vehicle
    - Petrol/diesel vehicle
    - Wood burning stove
    - Smart meters

    Create a graph to show your findings.
- In pairs, do some research to find out if there are any renewable energy projects underway in your local area. Create a short report describing the projects you have discovered.
- Choose a country in the EU and research their use of renewable energy.

Design

- Create a leaflet advising people on how to cut the costs of their energy bills during the energy crisis.
- Choose one of the articles in this book and create an illustration to highlight the key themes/message of your chosen article.
- Design a poster to promote one of the following unusual types of alternative energy sources:
    - Poo power
    - Sand batteries
    - Kinetic floor tiles
    - Smart solar flowers

Oral

- In small groups discuss what you think the three main reasons are for the current energy crisis. Share and compare your thoughts with other small groups in the class.
- There is much debate about reducing greenhouse gases and how renewable energy sources might form part of the answer. Most people agree that renewable energy is a good thing, but how do people react when a wind farm is proposed to be built near them? Role play a local council meeting with one group of students acting as councillors supporting the proposal, and another group of students acting as local residents who oppose it.
- Research the different energy suppliers in the UK and create a five minute presentation. What kind of different tariffs do they offer? Do they offer fixed price deals? How do they compare? Do the offer green/renewable energy options?
- In pairs, discuss the pros and cons of fossil fuels vs renewable energy.

Reading/Writing

- Read the article How much of the UK's energy is renewable? (page 8) and write a summary for your school/college newsletter.
- Write a letter to your school/college head to persuade them to use renewable energy in your school. Which types do you think will be best for your school/college? Use persuasive language to get your point across.
- Write an article explaining the causes of the energy crisis and what is meant by people having to choose between 'heating or eating'. Outline some ideas that you think could help to solve the crisis and the difficulties people are facing.

Index

A
air pollution 15, 37

anaerobic digestion 19

B
biofuels 2, 5, 9, 41

biogas energy 38

biomass 5, 38–39, 41

Boiler Upgrade Scheme 28

Bristol Energy Cooperative (BEC) 34–35

British Gas 21

C
carbon dioxide 8, 10–12

climate change 1, 3–4, 27, 36

coal 1, 2

   see also fossil fuels

cobalt production 15

cocoa plant matter 38–39

community energy 34–35

crematoriums, as energy source 39

D
decarbonization 10–12, 37

E
electric cars 11–12, 13–16, 27

electricity

   prices 7, 10–12, 37

   see also renewable energy

energy crisis 18–19, 20–28

energy management 25

energy prices 7, 10–12, 18–19, 20–23, 28, 37

energy savings 24–25

energy suppliers 22–23, 28

F
fossil fuels 1, 2–4, 8, 18–19, 28, 36, 37, 41

G
gas

   natural 2, 28

   prices 7, 10–12

geothermal energy 1–2, 5, 31, 41

green fluorescent protein (GFP) 39

Green Homes Grant 28

greenhouse gases 8, 36

greenwashing 41

H
heat pumps 2, 28, 31, 41

Hornsea 2 6

hydrogen power 33

hydropower 2, 5, 9, 31, 41

I
International Energy Agency (IEA) 30–31, 37

International Renewable Energy Agency (IRENA) 36

J
jellyfish 39

K
kinetic energy floors 38

L
Lewis, Martin 26

M
manure, as energy source 19

methane 39

MoneySavingExpert 26

N
net zero 8–9, 19, 23, 25, 38

nuclear power 29, 41

O

offshore wind farms 5, 6, 28, 31, 33, 41

Ofgem 7, 20, 22

oil, crude 2, 33

P

Paris Agreement 2016 1, 5

pollution 15, 37

R

renewable energy 1, 8–9, 19, 36–37

S

sand-powered batteries 32

seaweed 39

smart solar flowers 16–17

solar power 2, 5, 9, 11–12, 16–17, 30–31, 34–35, 37, 41

solar roads 39

T

tidal power 2

W

waste 37

   as energy source 2, 19, 38–39

wind power 2, 5–6, 9, 31, 33, 37, 41

Acknowledgements

The publisher is grateful for permission to reproduce the material in this book. While every care has been taken to trace and acknowledge copyright, the publisher tenders its apology for any accidental infringement or where copyright has proved untraceable. The publisher would be pleased to come to a suitable arrangement in any such case with the rightful owner.

The material reproduced in **issues** books is provided as an educational resource only. The views, opinions and information contained within reprinted material in **issues** books do not necessarily represent those of Independence Educational Publishers and its employees.

Images

Cover image courtesy of iStock. All other images courtesy Freepik, Pixabay & Unsplash.

Additional acknowledgements

With thanks to the Independence team: Shelley Baldry, Tracy Biram, Klaudia Sommer and Jackie Staines.

Danielle Lobban

Cambridge, September 2022